Practice Papers for SQA Exams

Higher

Biology
Practice Papers

© 2015 Leckie & Leckie Ltd
Cover © ink-tank

001/17022015

10 9 8 7 6 5 4 3

ISBN 9780007590926

Published by
Leckie & Leckie Ltd
An imprint of HarperCollins*Publishers*
Westerhill Road, Bishopbriggs, Glasgow, G64 2QT
T: 0844 576 8126   F: 0844 576 8131
leckieandleckie@harpercollins.co.uk   www.leckieandleckie.co.uk

Publisher: Peter Dennis
Project manager: Craig Balfour

Special thanks to
Donna Cole (copy edit)
Roda Morrison (proofread)
Louise Robb (proofread)
QBS (layout and illustration)

Printed and bound by CPI Group (UK) Ltd, Croydon, CR0 4YY

A CIP Catalogue record for this book is available from the British Library.

Acknowledgements

Images: P114 © Getty Images

Illustrations © HarperCollins Publishers

Introduction

Layout of the book

The four papers that follow have been produced to give you practice in two of the elements needed for Higher Grade Biology, reflecting Curriculum for Excellence's values and purposes. The first element is 'Knowledge and Understanding' [KU] which requires you to be familiar with all the theory. This is especially important in Biology, which is very much a knowledge-based subject: you need to know the facts, concepts and common techniques. The second element is 'Problem-Solving' [PS], which tests your ability to: select relevant information from a variety of sources; present information in a variety of forms; process information; plan, design and evaluate the design of experiments; draw valid conclusions and give explanations; and make predictions and generalisations. These two elements, which are assessed by the formal national exam, are combined to produce your overall grade. As a very rough guide, you will be awarded a pass at A, B, C if you obtain around 75%, 65% and 55% respectively.

All four papers have been designed to mirror the actual national exam. The layout and level of difficulty of all four papers are very closely modelled on the actual exam you will sit so you will become familiar with how it looks and how best to tackle it. In addition, the format of the questions reflects what you are likely to encounter in the exam. All parts of the syllabus have been represented in these papers.

Each paper has two sections:

- Section 1 consists of twenty multiple-choice questions, which test KU and PS. This section is worth 20 marks and is answered on a separate grid that is provided.

- Section 2 comprises restricted and extended response questions with a mixture of KU and PS. The majority of marks are for applying knowledge and understanding, with the other marks for applying scientific inquiry, scientific analytical thinking and problem-solving skills. This section is worth 80 marks and the questions are answered in spaces on the exam paper.

- The whole paper is worth 100 marks and you will have 2 hours and 30 minutes to complete it.

Pages 8–9 of this book have 'Links to the syllabus', which help you locate questions linked to particular sections of the syllabus or to one of the seven skills specified by the Scottish Qualifications Authority.

The answer section is at the back of the book. Each answer contains, where appropriate, some guidance as to how this was obtained, practical tips on how to tackle certain types of questions, details of how marks are awarded and advice on just what the examiners will be looking for. Also included is the type of question: knowledge and understanding or problem-solving.

How to use this book

Interacting with learning material is a powerful way to obtain feedback on your strengths and weaknesses. The material in these papers will give you an excellent way of working on different strategies needed to handle the Higher exam. Seeking help – where needed – from your teacher is vital to improvement and building up your confidence and expertise.

The papers can be used in two main ways:

1 You can complete an entire paper as preparation for the final exam. If you would like to use the book in this way, you can complete the paper under exam-style conditions by setting yourself a time for each paper and answering it as well as possible without using any references or notes. Alternatively, you can answer the paper questions as a revision exercise, using your notes to produce a model answer. Your teacher may mark these for you.

2 You can use the **Links to the syllabus** section on pages 8–9 to find all the questions within the book that deal with a specific topic in any of the three Course Units. This allows you to focus specifically on areas that you particularly want to revise or, if you are mid-way through your course, it lets you practise answering exam-style questions for just those topics that you have studied. You will find this particularly useful if you want practice on handling particular types of problem-solving questions.

Revision advice

The need to work out a plan for regular and methodical revision is obvious. If you leave things to the last minute, it may result in panic and stress which will inhibit you from performing to your maximum ability. If you need help, it is best to find this out when there is time to put it right. Revision planners are highly individual and you need to produce one that suits you. Use an area of your home that is set aside only for studying if possible, so that you form a positive link and in this way will be less liable to distractions. Not only do you need a plan for revising Higher Biology, but also for all your subjects. Below is one revision plan but you will have your own ideas here!

Work out a revision timetable for each week's work in advance – remember to cover all of your subjects and to leave time for homework and breaks. For example:

Day	6–6.45pm	7–8pm	8.15–9pm	9.15–10pm
Monday	Homework	Homework	English revision	Biology revision
Tuesday	Maths revision	Physics revision	Homework	Free
Wednesday	Geography revision	English revision	Biology revision	Maths revision
Thursday	Homework	Physics revision	Geography revision	Free
Friday	English revision	Biology revision	Free	Free
Saturday	Free	Free	Free	Free
Sunday	Maths revision	Physics revision	Geography revision	Homework

Make sure that you have at least one evening free each week to relax, socialise and re-charge your batteries. It also gives your brain a chance to process the information that you have been feeding it all week.

Arrange your study time into sessions that suit you, with a 15-minute break in between. Try to start studying as early as possible in the evening, when your brain is still alert, and be aware that the longer you put off starting the harder it will be.

If you miss a session, do not panic. Log this and make it up as soon as possible. Do not get behind in your schedule – discipline is everything in being a successful student.

Study a different subject in each session, except for the day before an exam.

Do something different during your breaks between study sessions – have a cup of tea, or listen to some music. Do not let your 15 minutes expand into 20 or 25 minutes!

Have your class notes and any textbooks available for your revision to hand, as well as plenty of blank paper, a pen, etc. If relevant, you may wish to have access to the Internet but be careful you restrict using this only for supporting your revision. You may also like to make keyword sheets like the example below:

Keyword	Meaning
Ribosome	Structure in the cell that manufactures protein
Fungicide	Chemical that kills fungi

Flashcards are another excellent way of practising terms and definitions. You can make these easily or buy them very cheaply. Use flashcards either to recall the keyword when you see the meaning or to give the meaning when you see the keyword. There are several websites that are free to use and give you the ability to generate flashcards online. If you collaborate with your friends and take different sections of the course, you can merge these into a very powerful learning and revision aid.

Finally, forget or ignore all or some of the advice in this section if you are happy with your present way of studying. Everyone revises differently, so find a way that works for you!

Command words

In the papers and in the Higher exam itself, a number of **command words** will be used in the questions. These command words are used to show you how you should answer a question: some words indicate that you should write more than others. If you familiarise yourself with these command words, it will help you to structure your answers more effectively.

Command word	Meaning/explanation
Name, state, identify, list	Giving a list is acceptable here – as a general rule you will get one mark for each point you give.
Suggest	Give more than a list – perhaps a proposal or an idea.
Outline	Give a brief description or overview of what you are talking about.
Describe	Give more detail than you would in an outline, and use examples where you can.
Explain	Discuss why an action has been taken or an outcome reached – what are the reasons and/or processes behind it?
Justify	Give reasons for your answer, stating why you have taken an action or reached a particular conclusion.
Define	Give the meaning of the term.
Compare/ contrast	Give the key features of **two** different items or ideas and discuss their similarities/differences.
Predict	Work out what will happen.

In the exam

Watch your time and pace yourself carefully. Work out roughly how much time you can spend on each answer and try to stick to this.

Be clear before the exam what the instructions are likely to be, for example how many questions you should answer in each section. The papers will help you to become familiar with the exam instructions.

Read the question thoroughly before you begin to answer it – make sure you know exactly what the question is asking you to do. If the question is in sections, for example, Paper B, Question 17, make sure that you can answer each section before you start writing.

Plan your extended responses by jotting down keywords, making a brief mind-map of the important points or whatever you find works best for you.

Do not repeat yourself as you will not get any more marks for saying the same thing twice. This also applies to annotated diagrams, which will not get you any extra marks if the information is repeated in the written part of your answer.

Give proper explanations. A common error is to give descriptions rather than explanations. If you are asked to explain something, you should be giving reasons. Check your answer to an **explain** question, and make sure that you have used linking words and phrases such as **because**, **this means that**, **therefore**, **so**, **so that**, **due to**, **since** and **the reason is**.

Good luck!

Links to the syllabus

Skill tested	Key area	S1 – Section 1 S2 – Section 2			
		Paper A	Paper B	Paper C	Paper D
Unit 1: DNA and the Genome *Demonstrating and applying knowledge*	The structure of DNA	S1: 2 S2: 1	S1: 1 S2: 1(a), (b) 2(a), (b), (c)	S1: 1 S2: 1(a), (b), (c)	S2: 1(a), (b)
	Replication of DNA	S1: 3 S2: 2	S1: 2 S2: 3(a), (b)	S1: 2 S2: 1(d), (e), (f)	S2: 1(c)
	Control of gene expression	S1: 4	S1: 4, 6 S2: 4A	S1: 3 S2: 5A	S1: 2 S2: 4
	Cellular differentiation	S2: 3	S2: 5	S1: 4 S2: 13B	S1: 3 S2: 5A
	The structure of the genome and mutations	S1: 5		S2: 3(a), (b), (c), (d)	S1: 4 S2: 8
	Evolution	S1: 7 S2: 4(a), (b)	S2: 7	S1: 5	S1: 5
	Genomic sequencing		S2: 17B	S2: 2(a), (b)(i), (ii), (c)	S1: 6 S2: 9
Unit 2: Metabolism and Survival *Demonstrating and applying knowledge*	Metabolism pathways and their control	S1: 8 S2: 5	S1: 7 S2: 8	S1: 7	S2: 5B, 10(a), (b), 11
	Cellular respiration	S1: 9	S1: 9 S2: 4B	S1: 8 S2: 6(a), (b), (c), (d), (e)	S1: 7 S2: 10(c), 13
	Metabolic rate	S1: 10	S1: 11	S1: 9 S2: 7(a), (b)	S1: 8 S2: 14
	Metabolism in conformers and regulators	S1: 11 S2: 6A	S2: 10	S1: 10	S1: 9
	Metabolism and adverse conditions	S2: 7	S1: 12 S2: 12	S1: 11 S2: 8(a), (b)	S1: 10
	Environmental control of metabolism	S1: 12		S1: 12 S2: 4(c), (d)	S1: 11 S2: 15
	Genetic control of metabolism	S2: 8(a), (c)			S1: 12
	Ethical considerations in the use of microorganisms, hazards and control of risks		S1: 13	S2: 13A	S1: 14

Skill tested	Key area	S1 – Section 1 / S2 – Section 2			
		Paper A	Paper B	Paper C	Paper D
Unit 3: Sustainability and Interdependence *Demonstrating and applying knowledge*	Food supply, plant growth and productivity	S1: 13,14 S2: 6B, 9(a), (b), (c)(ii)	S1: 14 S2: 13		S1: 15 S2: 6,16A
	Plant and animal breeding	S1: 16	S1: 15 S2: 14	S2: 9(c)	S1: 18
	Crop protection	S1: 17 S2: 10(c)	S1: 16 S2: 15	S1: 15 S2: 11(a), (b), (c), (d)	
	Animal welfare	S2: 11(a), (b), (c)	S1: 17	S1: 16 S2: 12(a)	S2: 17
	Symbiosis	S2: 14A	S1: 18	S1: 17	S2: 18
	Social behaviour	S1: 18 S2: 11(d), (e)	S2: 17A	S2: 5B	S1: 19
	Mass extinction and biodiversity	S2: 14B(i)		S1: 18 S2: 10(a), (d)	S2: 16B
	Threats to biodiversity	S1: 19 S2: 14B(ii)	S1: 20		
Higher Biology Course *Skills of scientific enquiry*	Planning investigations	S2: 8(b)(ii), 9(c)(i), 12(a), (d)	S1: 3 S2: 1(c), 3(d), 16(c)	S1: 19 S2: 9(a), (b)	S1: 17 S2: 3, 7(b)
	Selecting information	S1: 20 S2: 4(a)(ii), (b)(ii), 8(b)(i), 13(a)	S1: 19 S2: 6(a), (b), 11(a)	S1: 6 S2: 7(e)	S2: 12(b)
	Presenting information	S2: 12(e)	S2: 6(c), 16(a)	S2: 4(e)	S2: 2(a)
	Processing information	S1: 1, 6 S2: 10(a), 12(g), 13(b)	S1: 5,10 S2: 2(d), (e), 3(c), 9(a), (b)	S1: 14, 20 S2: 4(b), 7(d), 9(d), 12(b), (c)	S1: 16 S2: 2(c), 12(a)
	Predicting and generalising	S2: 12(h)		S2: 10(c)	S1: 20 S2: 7(d)
	Concluding and explaining	S1: 15	S2: 10(b), 12(f)	S1: 8	S2: 9(b), 16
	Evaluating		S2: 8b(iii), 12(b), (c)		S2: 6(b), 11(c)

SECTION 1 ANSWER GRID

Mark the correct answer as shown ◉

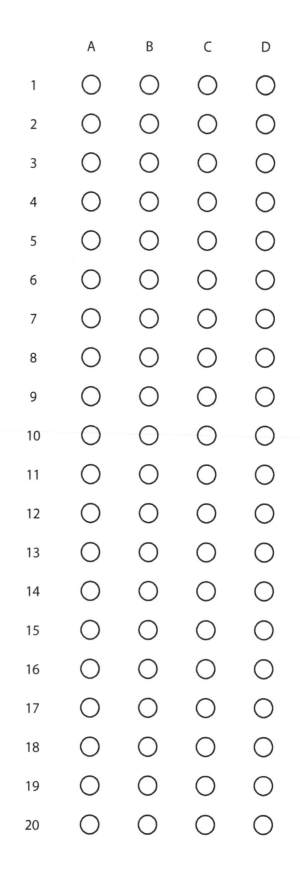

	A	B	C	D
1	○	○	○	○
2	○	○	○	○
3	○	○	○	○
4	○	○	○	○
5	○	○	○	○
6	○	○	○	○
7	○	○	○	○
8	○	○	○	○
9	○	○	○	○
10	○	○	○	○
11	○	○	○	○
12	○	○	○	○
13	○	○	○	○
14	○	○	○	○
15	○	○	○	○
16	○	○	○	○
17	○	○	○	○
18	○	○	○	○
19	○	○	○	○
20	○	○	○	○

Paper A

CfE Higher Biology

Practice Papers for SQA Exams

Paper A

Fill in these boxes and read what is printed below.

Full name of centre

Town

Forename(s)

Surname

Answer all of the questions in the time allowed.

Total marks – 100

Section 1 – 20 marks

Section 2 – 80 marks

Read all questions carefully before attempting.

You have 2 hours 30 minutes to complete this paper.

Write your answers in the spaces provided, including all of your working.

Leckie×Leckie

Scotland's leading educational publishers

SECTION 1 – 20 marks

Attempt ALL questions

Answers should be given on the separate answer sheet provided.

1. A length of double-stranded DNA is 250 base pairs long. Of these bases, 60 are cytosine.

    The percentage of thymine bases present in this part of the DNA is

    A 12%

    B 19%

    C 24%

    D 38%

2. In which form does the genetic material of a eukaryote exist within the nucleus?

    A Linear chromosomes

    B Circular chromosomes

    C Linear plasmids

    D Circular plasmids.

3. The following are steps in the process of polymerase chain reaction (PCR).

    1 Cooling allows primers to bind to target sequences

    2 DNA is heated to separate the strands

    3 Repeated cycles of heating and cooling amplify this region of DNA

    4 Heat-tolerant DNA polymerase then replicates the region of DNA.

    Choose the letter that puts the above steps into the correct order.

    A $1 \rightarrow 2 \rightarrow 3 \rightarrow 4$

    B $2 \rightarrow 1 \rightarrow 3 \rightarrow 4$

    C $2 \rightarrow 1 \rightarrow 4 \rightarrow 3$

    D $4 \rightarrow 2 \rightarrow 3 \rightarrow 1$

4. During transcription, which regions of a primary transcript are non-coding and removed in RNA splicing?

A Codons

B Anti-codons

C Introns

D Exons.

5. The following sequence of bases codes for three amino acids.

T–G–C–A–A–G–C–G–T

The sequence of bases is altered by a mutation, and is changed to the sequence below.

T–G–C–A–A–C–C–G–T

Which type of mutation has occurred?

A Deletion

B Insertion

C Substitution

D Duplication.

6. The graph on the right shows the rate of photosynthesis at two different levels of carbon dioxide concentration at 20 °C.

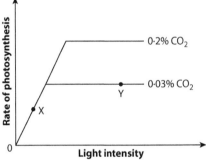

From the evidence given, which line in the table below identifies the factors most likely to be limiting the rate of photosynthesis at points **X** and **Y** on the graph?

	X	**Y**
A	light intensity	CO_2 concentration
B	temperature	light intensity
C	CO_2 concentration	temperature
D	light intensity	temperature

7. Which of the following terms describes the generation of new biological species by evolution?

A Selection

B Speciation

C Gene transfer

D Genetic drift.

8. Which type of enzyme inhibition involves the binding of the inhibitor to the active site?

A Competitive

B Non-competitive

C Feedback

D End product.

9. Which respiratory substrate is used during prolonged starvation when other reserves are exhausted?

A Starch

B Glucose

C Tissue protein

D Fat.

10. The following diagram shows part of a metabolic pathway.

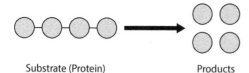

Substrate (Protein)   Products

Which line in the table shows the type of reaction shown above and the products?

	Reaction type	*Products*
A	anabolic	amino acids
B	catabolic	amino acids
C	anabolic	glucose
D	catabolic	glucose

11. Some species of fish can migrate from salt to freshwater habitats every year with their breeding cycle. They are regulators.

Which of the following statements about regulators is **true**?

A They do not require energy to achieve homeostasis

B They control their internal environment, which increases the range of possible ecological niches

C Their internal environment is dependent upon their external environment

D They may have low metabolic costs and a narrow ecological niche.

12. During the growth of unicellular organisms, in which phase do the cells grow and multiply at the maximum rate?

A Lag

B Exponential

C Stationary

D Death.

13. The diagram below represents part of the Calvin cycle within a chloroplast.

Which line in the table below shows the effect of decreasing CO_2 availability on the concentrations of RuBP and GP in the cycle?

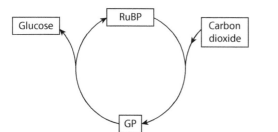

	RuBP concentration	**GP concentration**
A	increase	decrease
B	decrease	increase
C	increase	increase
D	decrease	decrease

14. Which of the following is **not** a factor affecting food security?

A Quantity

B Access

C Quality

D Abiotic.

15. The graph below shows the changes in the populations of red and grey squirrels in an area of woodland over a 10-year period.

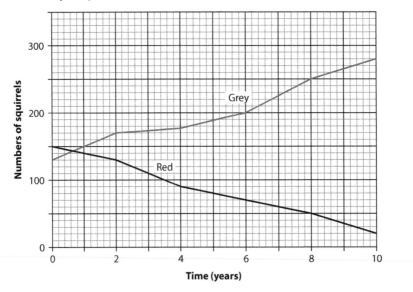

From the graph the following conclusions were suggested.

1 The grey squirrel population increased by 150% over the 10-year period.

2 The red squirrel numbers decreased from 150 to 20 over the 10-year period.

3 After 8 years the grey squirrel population was five times greater than the red.

Which of the conclusions are **correct**?

A 1 and 2 only

B 1 and 3 only

C 2 and 3 only

D 1, 2 and 3

16. Which of the following statements about outbreeding is **true**?

A Can lead to a loss of heterozygosity

B Involves the fusion of two gametes from unrelated members of the same species

C Involves the fusion of two gametes from close relatives

D Is naturally occurring in some species of self-pollinating plants.

17. Which of the following is **not** an ideal characteristic of a pesticide?

A Soluble

B Specific

C Short-lived

D Safe.

18. Which of the following would be considered appeasement behaviour in primates?

A Roaring

B Standing tall

C Chest beating

D Grooming.

19. Which term describes a species that humans have moved either intentionally or accidentally to new geographic locations?

A Introduced

B Naturalised

C Invasive

D Native.

20. The graph below shows changes that occur in the masses of protein, fat and carbohydrate in the body of a hibernating mammal during 7 weeks without food.

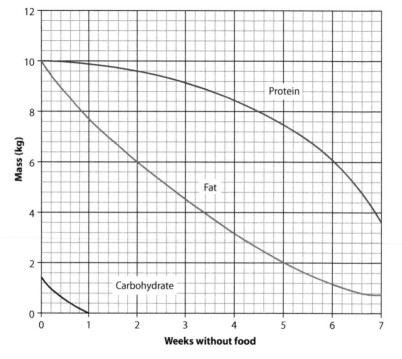

What percentage of the original mass of fat was used up between weeks 2 and 5?

A 33%

B 40%

C 67%

D 80%

MARKS

SECTION 2 – 80 marks

Attempt ALL questions

It should be noted that questions 6 and 14 contain a choice.

1. Below are two types of cell with different DNA organisation.

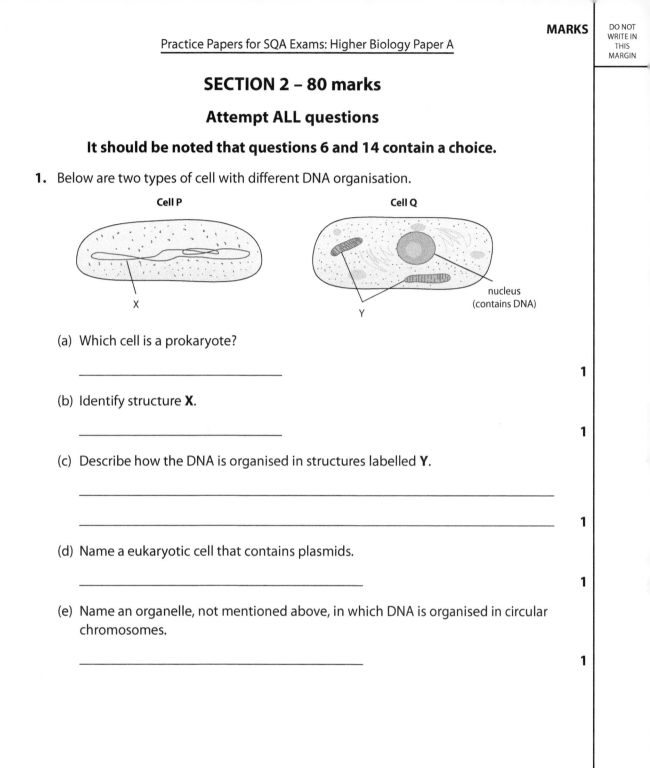

Cell P

Cell Q

X

Y

nucleus
(contains DNA)

(a) Which cell is a prokaryote?

_____   1

(b) Identify structure **X**.

_____   1

(c) Describe how the DNA is organised in structures labelled **Y**.

_____

_____   1

(d) Name a eukaryotic cell that contains plasmids.

_____   1

(e) Name an organelle, not mentioned above, in which DNA is organised in circular chromosomes.

_____   1

MARKS

2. The diagram below shows the first stage of protein synthesis in a cell.

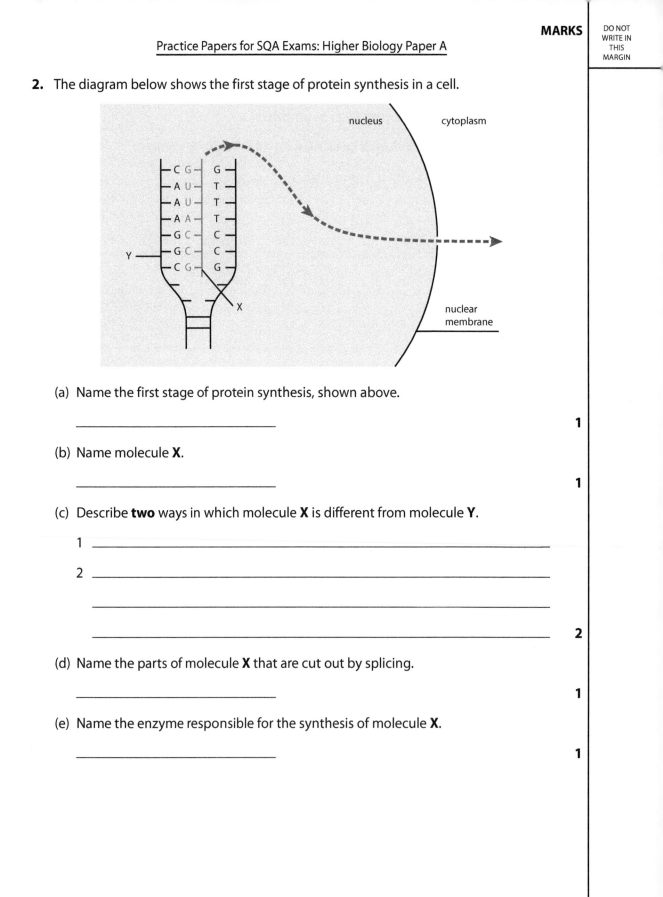

(a) Name the first stage of protein synthesis, shown above.

_____ **1**

(b) Name molecule **X**.

_____ **1**

(c) Describe **two** ways in which molecule **X** is different from molecule **Y**.

1 _____

2 _____

_____

_____ **2**

(d) Name the parts of molecule **X** that are cut out by splicing.

_____ **1**

(e) Name the enzyme responsible for the synthesis of molecule **X**.

_____ **1**

MARKS

DO NOT
WRITE IN
THIS
MARGIN

Practice Papers for SQA Exams: Higher Biology Paper A

3. Stem cells are unspecialised somatic cells in animals.

(a) State **two** characteristics of stem cells.

1 _____

_____

2 _____

_____ **2**

(b) Name the type of stem cell capable of differentiating into all types of cell.

_____ **1**

(c) Give an example of a medical condition that, in the future, may be treated using stem cells.

_____ **1**

(d) Give **one** example of an ethical issue that using stem cells raises.

_____

_____ **1**

MARKS

4. Evolutionary relatedness among different groups of organisms can be studied and constructed, like in the diagram below.

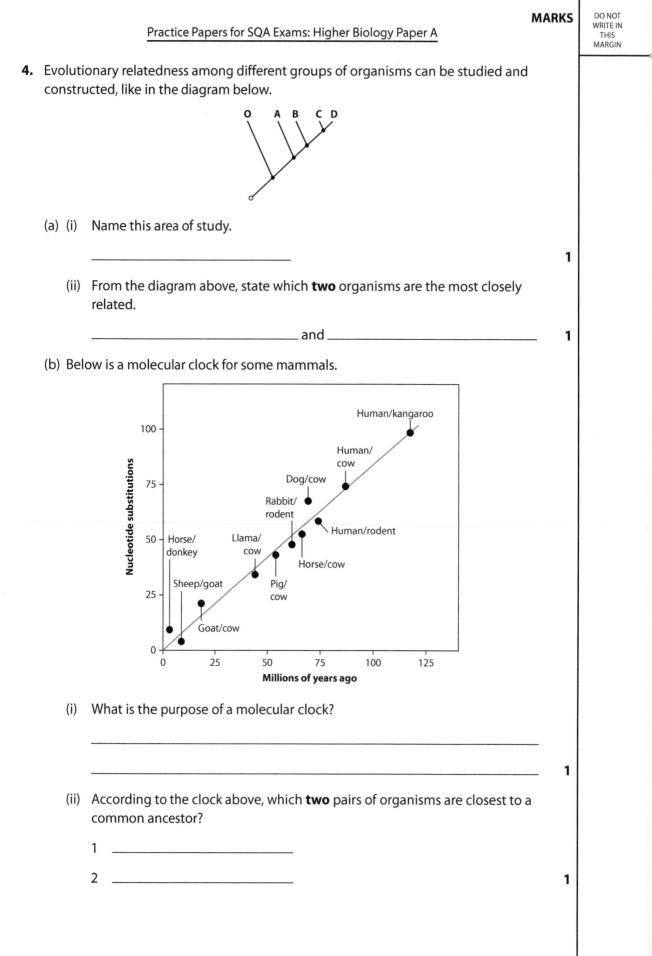

(a) (i)   Name this area of study.

_____   1

(ii) From the diagram above, state which **two** organisms are the most closely related.

_____ and _____   1

(b) Below is a molecular clock for some mammals.

(i)   What is the purpose of a molecular clock?

_____

_____   1

(ii) According to the clock above, which **two** pairs of organisms are closest to a common ancestor?

1   _____

2   _____   1

5. The diagram below shows an enzyme, its substrate and the enzyme-substrate complex.

Substrate   Enzyme-Substrate Complex

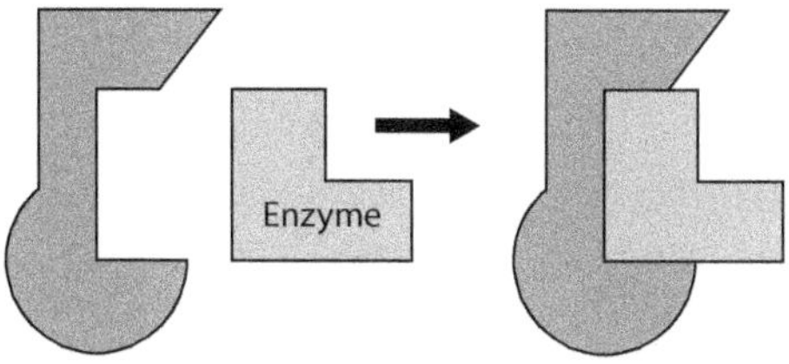

(a) Where does a substrate bind to an enzyme?

_____

1

(b) Give **two** properties of an enzyme.

1 _____

2 _____

2

(c) Enzymes are biological catalysts.

State **three** properties or functions of catalysts.

1 _____

2 _____

3 _____

3

MARKS

6. Answer **either A or B**.

   A Give an account of metabolism in conformers and regulators.   **4**

   OR

   B Give an account of the Calvin cycle.   **4**

   Labelled diagrams may be used where appropriate.

7. Organisms can regulate their internal environment using the control mechanism below.

(a) State the general name given to the control mechanism shown.

_____ **1**

(b) Animals can be divided into two categories, depending on their ability to regulate their body temperatures.

Name each category and give **one** example of an organism for each.

1 Name _____ Example _____ **2**

2 Name _____ Example _____ **2**

(c) The hypothalamus is the temperature-monitoring centre in mammals.

Explain how the hypothalamus detects changes in temperature of the blood and how it responds to maintain thermoregulation.

Detects _____

Responds _____

_____ **2**

MARKS

8. (a) Decide if each of the statements relating to genetic control of metabolism in the table below is true or false and tick (✓) the appropriate box.

If you decide that the statement is false, write the correct term in the correction box to replace the term underlined in the statement.

Statement	True	False	Correction
Exposure to UV light may result in <u>mutations</u>, some of which may produce an improved strain.			
Some <u>fungi</u> can transfer plasmids or pieces of chromosomal DNA to each other or take up DNA from their environment to produce new strains.			
In fungi and yeast, new phenotypes can be brought about by <u>asexual</u> reproduction between existing strains.			

2

(b) The table below shows the results of an experiment where two different yeast cultures are exposed to UV light for 24 hours. Two cultures, A and B, are grown into ten colonies before the start of the experiment.

Exposure time to UV light (hours)	Number of colonies	
	Yeast culture A	Yeast culture B
0	10	10
6	8	10
12	5	10
18	2	10
24	0	10

(i) Which yeast culture – **A** or **B** – is most sensitive to UV light?

_____

1

(ii) Give **two** variables that would have to be kept constant during this experiment.

Variable 1 _____

Variable 2 _____

2

(iii) State **one** way in which the accuracy of these results could be improved.

_____

_____

1

MARKS

DO NOT
WRITE IN
THIS
MARGIN

Practice Papers for SQA Exams: Higher Biology Paper A

(c) UV light is a mutagenic agent.

   Give an example of another mutagenic agent.

_____ **1**

9. The graph below shows the absorption spectra of three different leaf pigments.

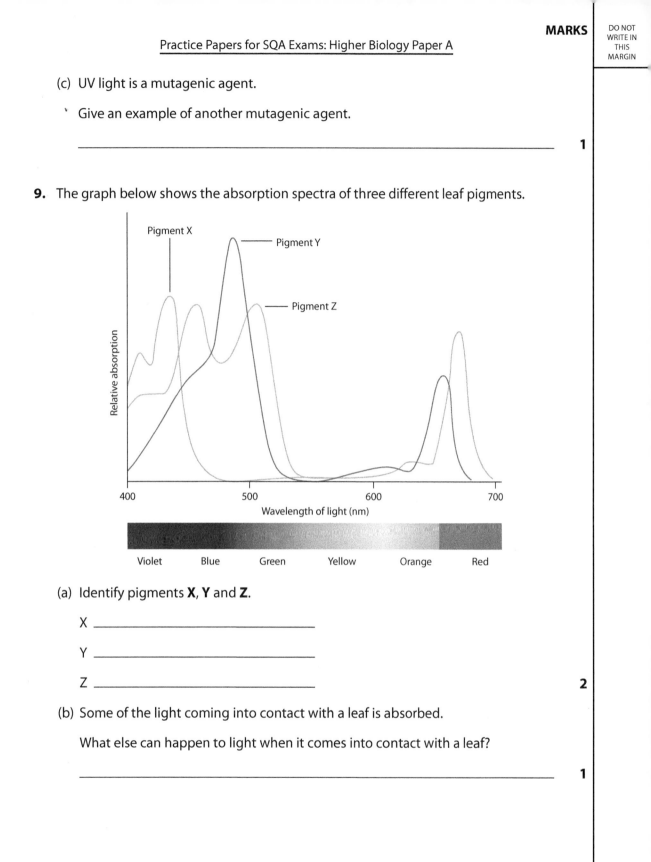

(a) Identify pigments **X**, **Y** and **Z**.

   X _____

   Y _____

   Z _____ **2**

(b) Some of the light coming into contact with a leaf is absorbed.

   What else can happen to light when it comes into contact with a leaf?

_____ **1**

(c) An experiment was set up to determine the rate of photosynthesis, in *Elodea,* using different wavelengths of light. The number of bubbles produced per minute were counted.

The apparatus is shown below.

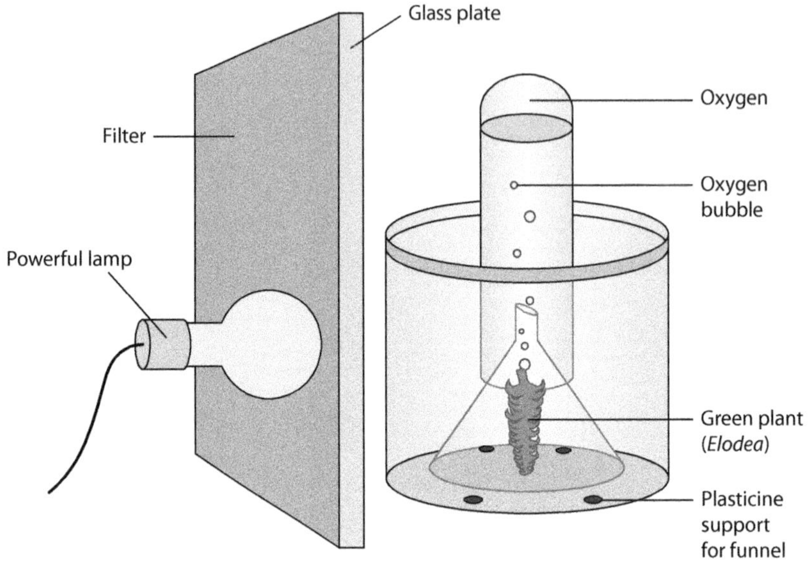

(i) Describe the purpose of the glass plate in this experiment.

_____

_____ **1**

(ii) The results from this experiment could be graphed to show an action spectrum, a measure of the rate of photosynthesis.

State the colour of filter that would produce the greatest rate of photosynthesis.

_____ **1**

10. The graph below shows how world population and fertiliser usage changed between 1920 and 2000.

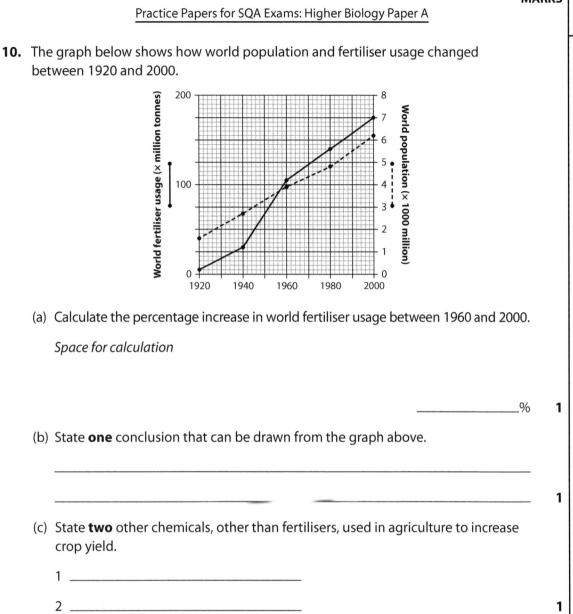

(a) Calculate the percentage increase in world fertiliser usage between 1960 and 2000.

Space for calculation

_____% **1**

(b) State **one** conclusion that can be drawn from the graph above.

_____

_____ **1**

(c) State **two** other chemicals, other than fertilisers, used in agriculture to increase crop yield.

1 _____

2 _____ **1**

MARKS

DO NOT
WRITE IN
THIS
MARGIN

Practice Papers for SQA Exams: Higher Biology Paper A

11. The behaviours of 10 monkeys in an enclosed area were observed for 24 hours.

   (a) What name is given to the study of animal behaviour?

   _____   **1**

   (b) Some monkeys were observed chewing their own tail.

   What term is used to describe this type of behaviour?

   _____   **1**

   (c) Describe **one** way in which this type of behaviour could be reduced in the monkeys.

   _____

   _____   **1**

   (d) Give **two** examples of behaviour used by primates to reduce conflict.

   1 _____   **1**

   2 _____   **1**

   (e) One of the smallest monkeys developed an alliance with the dominant monkey.

   State **one** advantage of this alliance.

   _____

   _____   **1**

MARKS

12. An investigation into the metabolic rate of a stick insect at rest, using the apparatus shown in the diagram below, measured its oxygen uptake.

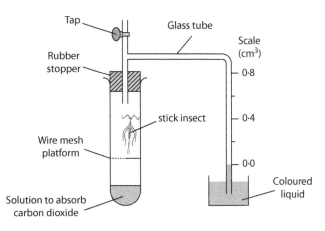

To start the experiment, the tap was closed and the reading on the scale recorded.

Every 2 minutes, for 10 minutes, readings from the scale were recorded.

The apparatus was kept at 15 °C with the tap open for 10 minutes.

The results are shown in the table below.

Time after tap closed (minutes)	Oxygen uptake (cm³)
0	0·00
2	0·08
4	0·16
6	0·24
8	0·36
10	0·40

(a) Identify the dependent variable in this experiment.

_____

1

(b) Describe how the experimental procedure could be improved to increase the reliability of the results.

_____

1

(c) An identical apparatus was set up without the stick insect as a control.

Explain why the use of a control ensures valid results.

_____

_____

1

MARKS

DO NOT
WRITE IN
THIS
MARGIN

Practice Papers for SQA Exams: Higher Biology Paper A

(d) Explain why the apparatus was left for 10 minutes with the tap open before readings were taken.

_____

_____ **1**

(e) On the grid below, plot a line graph to show the oxygen uptake against time.

2

(f) From the results of this investigation, describe the changes in oxygen uptake of the stick insect.

_____

_____

_____ **2**

(g) The mass of the stick insect was 10·0 g.

Use the results in the table to calculate the average rate of oxygen uptake per gram of stick insect per minute over the 10 minute period.

Space for calculation

_____ cm^3 per gram per minute **1**

(h) Predict the effect of a decrease in temperature from 15 °C to 5 °C on the oxygen uptake by the stick insect and justify your answer.

Prediction _____

Justification _____

_____ **1**

13. The graph below shows the annual consumption of fertiliser in major regions of the world.

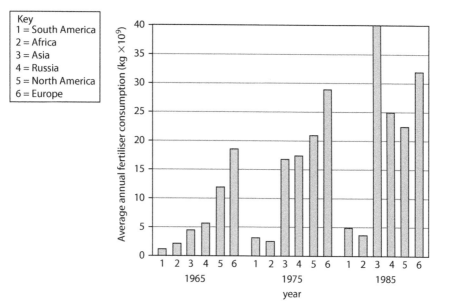

Key
1 = South America
2 = Africa
3 = Asia
4 = Russia
5 = North America
6 = Europe

(a) Using values from the graph, describe the changes in the use of fertiliser in Africa and Asia in 1965 compared with 1985.

_____

_____

_____ **2**

(b) Calculate the yearly increase in average annual fertiliser consumption in Europe between 1965 and 1985.

Space for calculation

_____ kg $\times 10^9$ per year   **1**

MARKS

14. Answer **either A or B** in the space below.

   A Describe symbiosis under the following headings.   **9**

   (i) Parasitism

   (ii) Mutualism

   OR

   B Describe biodiversity under the following headings.   **9**

   (i) Measuring biodiversity

   (ii) Threats to biodiversity

   Labelled diagrams may be used where appropriate.

[END OF QUESTION PAPER]

SECTION 1 ANSWER GRID

Mark the correct answer as shown ◉

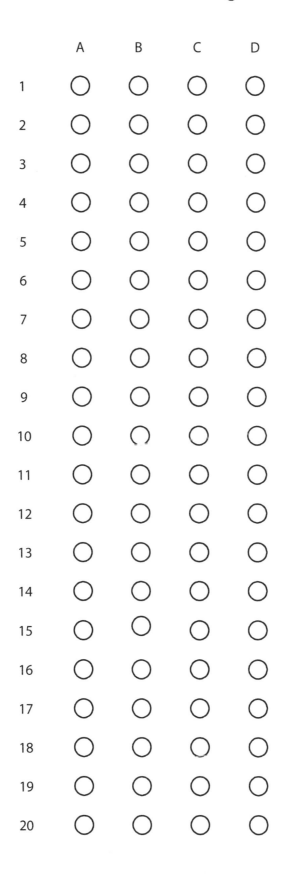

Paper B

CfE Higher Biology

Practice Papers for SQA Exams

Paper B

Fill in these boxes and read what is printed below.

Full name of centre

Town

Forename(s)

Surname

Answer all of the questions in the time allowed.

Total marks – 100

Section 1 – 20 marks

Section 2 – 80 marks

Read all questions carefully before attempting.

You have 2 hours 30 minutes to complete this paper.

Write your answers in the spaces provided, including all of your working.

Scotland's leading educational publishers

SECTION 1 – 20 marks

Attempt ALL questions

Answers should be given on the separate answer sheet provided.

1. Which line in the table correctly shows where deoxyribose and phosphate are found in each strand of DNA?

	3´ end	5´ end
A	deoxyribose	phosphate
B	phosphate	deoxyribose
C	deoxyribose and phosphate	neither present
D	neither present	deoxyribose and phosphate

2. DNA polymerase

    A causes the formation of hydrogen bonds between bases

    B cannot add nucleotides to an already existing DNA chain

    C requires a primer to be present

    D can only add nucleotides to the free 5´ end of a DNA strand.

3. A student investigated the effect on growth of different volumes of water added to three groups of cloned tobacco plants over a 1-week period. Each group contained the same number of plants. The first group received 300 cm³ water; the second group received 200 cm³ water and the third group received 100 cm³ water.

    What would be a suitable control for this experiment?

    A Increase the size of all the groups

    B Use tobacco plants which were not clones

    C Use a fourth group of plants to which 50 cm³ water was added

    D Use a fourth group of plants which had no water added at all.

4. Gene expression is controlled by regulating

A translation only

B transcription only

C both translation and transcription

D post-translational modification.

5. A double-stranded DNA fragment was found to have 180 nucleotides present. Which combination of bases would fit into this fragment?

A 40 thymine and 50 adenine

B 50 adenine and 50 cytosine

C 40 cytosine and 50 guanine

D 50 cytosine and 40 adenine.

6. Introns are

A sections of DNA that code for mRNA

B enzymes that join up DNA fragments

C non-coding lengths of DNA

D segments of DNA that repeat themselves.

7. Which of the following proteins combines with sodium ions and helps move them across the cell membrane?

A Carrier

B Channel

C Receptor

D Enzyme.

8. The concentration of ions in a cell from a plant that lives in salt water are shown in the following table, along with their concentrations in the water surrounding it.

	Concentration of ions [mg/litre]		
	Chloride	Potassium	Sodium
Cell contents	0·50	0·05	0·50
Salt water	0·55	0·50	0·10

Which of the following is a valid conclusion from these data?

The cells of this plant

A selectively concentrate chloride ions

B actively pump out potassium ions into the salt water

C actively pump out sodium ions into the salt water

D cannot distinguish between chloride and potassium ions.

9. During aerobic respiration, NADH is not produced in

A glycolysis

B the electron transport chain

C the Krebs cycle

D pyruvate oxidation.

10. A human nerve cell was magnified 80 000 times and measured 260 mm.

The length of the cell in micrometres is

A 1·25

B 1·50

C 3·00

D 3·25

11. Which line in the table below correctly describes features of the circulatory system found in a mammal?

	Features	
	Heart	**Circulation**
A	four chambers	incomplete double
B	septum present	complete double
C	four chambers	complete single
D	septum absent	complete double

12. To survive very cold conditions an animal, for example an Arctic fox, would have

A a large surface area to volume ratio

B large ears with a dense blood supply

C a round and compact body shape

D a very thin layer of fat underneath the skin.

13. A new, microbially-based medical product has come on the market.

Which of the following would **not** be an ethical consideration in its development?

A Microorganisms do not feel pain as we understand this

B More product can be produced by genetically modified microbes than by nature

C Some strains of the live microbe could escape into the environment

D The product would be free from impurities.

14. Which of the following would be most likely to contribute to increasing food security in an environmentally friendly way?

A Transferring genetic material from domestic strains into wild strains

B Extending the use of insecticides

C Using more organically-based fertilisers

D Decreased plant productivity.

15. Which of the following statements about breeding is correct?

    A Inbreeding involves the fusion of two gametes from close relatives

    B Inbreeding involves the fusion of two gametes from different species

    C Outbreeding involves the fusion of gametes from different species

    D Continual outbreeding leads to loss of heterozygosity in a species.

16. Which of the following characteristics do perennial weeds usually possess?

    A Fast-growing and long life-cycles

    B Slow-growing and produce dormant seeds

    C Possess storage organs and can reproduce asexually

    D Produce few seeds and cannot reproduce asexually.

17. Animal welfare ensures animals

    A that are used in the human food chain are killed humanely

    B do not exhibit natural behaviour patterns

    C are kept in safe conditions for part of the year

    D are reared in isolation to prevent infection.

18. Which of the following feeding relationships is classed as mutualism?

    A Lice on human hair

    B Cellulose-digesting bacteria in the gut of cattle

    C Leech sucking blood from fish

    D Tapeworm living inside dog.

19. The following pie-chart shows causes of death in a Scottish city for males in 2012.

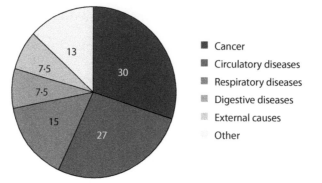

Which of the following statements is true, based on these data?

A Cancer causes three times as many deaths as respiratory and digestive diseases combined

B Deaths due to other causes number approximately half as many as those caused by circulatory diseases

C 40% of deaths are due to cancer and respiratory diseases combined

D External causes are responsible for twice as many deaths compared to digestive diseases.

20. Which of the following would be most likely to have a negative effect on biodiversity?

A Planting new trees along a motorway

B Leaving areas of land unmanaged in a woodland

C Connecting isolated habitats that were previously in contact

D Introducing an invasive species into an environment.

SECTION 2 – 80 marks

Attempt ALL questions

It should be noted that questions 4 and 17 contain a choice.

1. An early experiment that helped establish that DNA could be transferred from a dead cell to a live cell, involved the use of a bacterium. This bacterium exists in two forms: smooth (S) and rough (R). One, 'strain S', produces pneumonia in mice, causing death; and the other, 'strain R', is harmless to mice. This bacterium can be killed by heat.

The diagram below shows how the experiment was carried out.

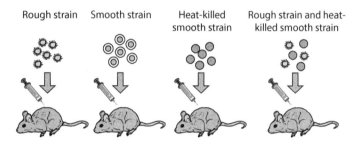

Rough strain   Smooth strain   Heat-killed smooth strain   Rough strain and heat-killed smooth strain

(a) Predict what will happen to each of the four mice after being injected as shown.

Mouse 1 _____

Mouse 2 _____

Mouse 3 _____

Mouse 4 _____   **2**

(b) Blood samples were taken from each mouse. Which sample(s) would contain live S bacteria?

_____   **1**

(c) Which mice formed controls for this experiment?

_____   **1**

2. (a) The diagram below represents a strand of DNA.

5′ AACGGGTTTATGCGT 3′

Write out the complementary strand in the space below.

1

(b) Give the feature of a DNA molecule that determines its melting point.

_____

_____ 1

(c) A single strand of DNA is much longer than the nucleus in which it is found. Describe how the molecule is arranged so that one strand does not get mixed up with other strands in the nucleus.

_____

_____ 2

(d) A chromosome in an animal cell was estimated to be 40 mm long. However, in the nucleus, its length was apparently reduced by 200 000 times. What would this apparent length be in micrometres?

Space for calculation

_____ micrometres 1

(e) The mass of a cell's genome is measured in picograms (pg).

1 picogram = 10^{-12} gram

To convert the mass of the genome to a number of base pairs, a formula is used, as follows:

number of base pairs = mass (pg) × 0·978 × 10^9

Calculate the mass (pg) of a chromosome that has 58 680 000 base pairs.

Space for calculation

_____ pg 1

MARKS

DO NOT
WRITE IN
THIS
MARGIN

Practice Papers for SQA Exams: Higher Biology Paper B

3. The polymerase chain reaction (PCR) can be used to produce many copies of a piece of DNA in the laboratory. The flow chart below shows how a sample of DNA was treated during one cycle.

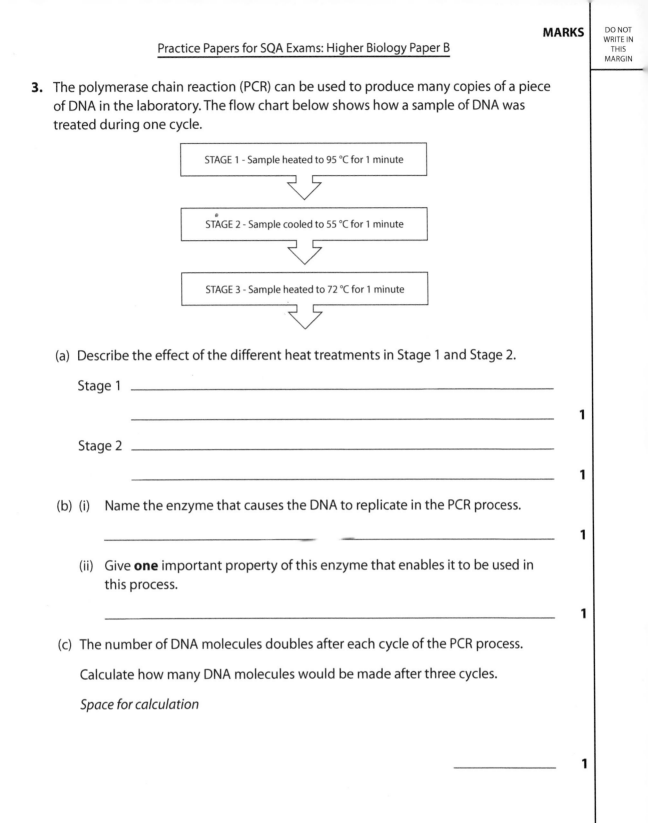

STAGE 1 - Sample heated to 95 °C for 1 minute

STAGE 2 - Sample cooled to 55 °C for 1 minute

STAGE 3 - Sample heated to 72 °C for 1 minute

(a) Describe the effect of the different heat treatments in Stage 1 and Stage 2.

Stage 1 _____

_____ **1**

Stage 2 _____

_____ **1**

(b) (i) Name the enzyme that causes the DNA to replicate in the PCR process.

_____ **1**

(ii) Give **one** important property of this enzyme that enables it to be used in this process.

_____ **1**

(c) The number of DNA molecules doubles after each cycle of the PCR process.

Calculate how many DNA molecules would be made after three cycles.

Space for calculation

_____ **1**

(d) The tube below shows the mixture of chemicals used in the PCR process.

Enzyme and buffer
Primers
DNA nucleotides
DNA

What would be found in a control tube designed to show the need for enzyme to be present?

_____

_____ **1**

MARKS

DO NOT
WRITE IN
THIS
MARGIN

4. Answer **either A or B**.

A Describe the structure of RNA. 4

OR

B Describe how fats and proteins can be used as respiratory substrates. 4

Labelled diagrams may be used where appropriate.

MARKS

5. (a) The diagram below shows a part of a root of a flowering plant.

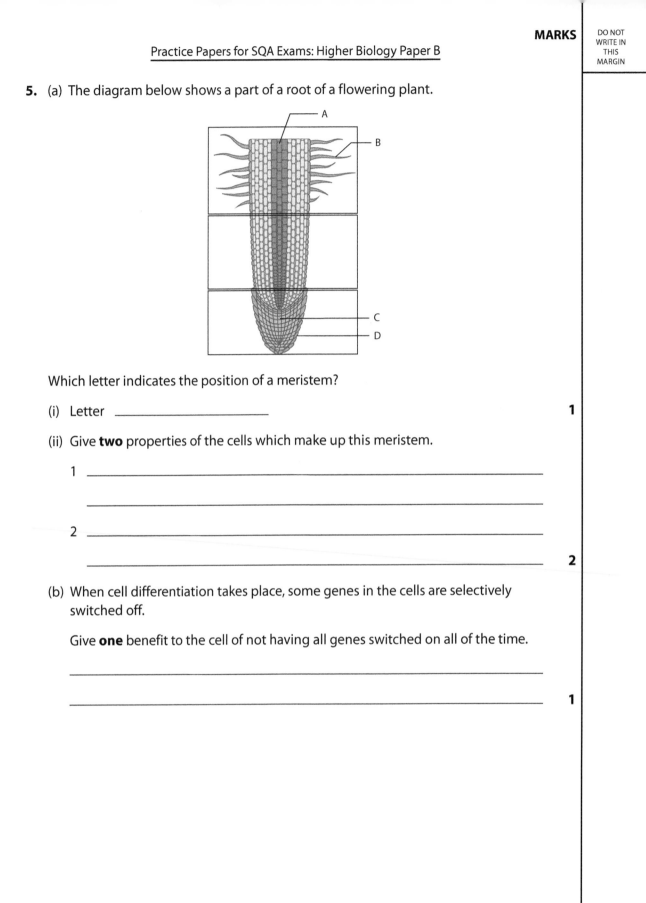

Which letter indicates the position of a meristem?

(i) Letter _____   **1**

(ii) Give **two** properties of the cells which make up this meristem.

1 _____

_____

2 _____

_____   **2**

(b) When cell differentiation takes place, some genes in the cells are selectively switched off.

Give **one** benefit to the cell of not having all genes switched on all of the time.

_____

_____   **1**

MARKS

DO NOT
WRITE IN
THIS
MARGIN

Practice Papers for SQA Exams: Higher Biology Paper B

6. The graph below shows the results of an experiment into how changing the light intensity and concentration of carbon dioxide affects the rate of photosynthesis in a plant.

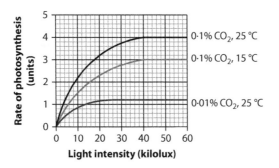

(a) At a light intensity of 50 kilolux, which factor, as shown in the graph, has the greater effect in increasing the rate of photosynthesis?

Justify your answer.

Factor _____

Justification _____

_____

_____ **1**

(b) What combination of light intensity, carbon dioxide concentration and temperature gives the maximum rate of photosynthesis?

_____ **1**

(c) The experiment was repeated at a carbon dioxide concentration of 0·01% and a temperature of 20 °C. Draw, **on the grid above**, a curve to show what the predicted results would be. **1**

(d) The results shown were based on one run of the experiment. How could the reliability of the results be improved?

_____ **1**

7. The phylogenetic tree below shows some of the evolutionary relationships between different animals.

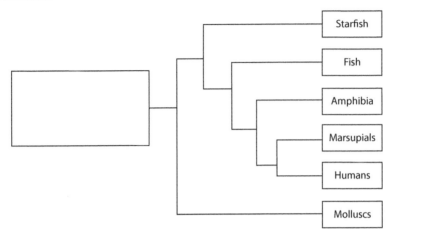

(a) All these animals evolved from one shared earlier form. What is this earlier form called? Write your answer into the blank box.

1

(b) Humans are more closely related to marsupials than to amphibia.

Using the information on the tree, explain how this is known.

_____

_____ **1**

(c) Give the group of animals to which molluscs are most closely related.

Use information from the phylogenetic tree to explain your answer.

Group of animals _____

Explanation _____

_____ **1**

MARKS

DO NOT
WRITE IN
THIS
MARGIN

Practice Papers for SQA Exams: Higher Biology Paper B

8. (a) The following diagram shows two different types of reaction that can occur in cells.

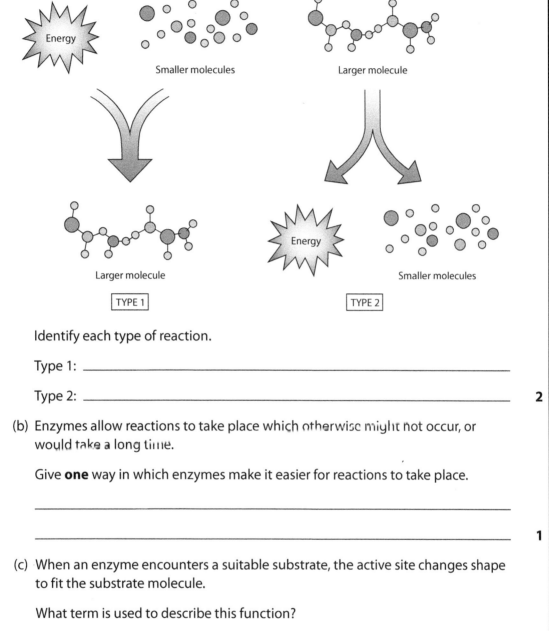

Identify each type of reaction.

Type 1: _____

Type 2: _____   **2**

(b) Enzymes allow reactions to take place which otherwise might not occur, or would take a long time.

Give **one** way in which enzymes make it easier for reactions to take place.

_____

_____   **1**

(c) When an enzyme encounters a suitable substrate, the active site changes shape to fit the substrate molecule.

What term is used to describe this function?

_____   **1**

MARKS

(d) When an end product of an enzyme-catalysed reaction accumulates, it may cause the activity of the enzyme involved to decrease or stop temporarily.

What term is used to describe this type of control?

_____   **1**

(e) Some enzyme control is brought about by the action of inhibitor molecules which have a shape very similar to the enzyme's normal substrate and so can bind to the active site.

What type of inhibition is this?

_____   **1**

MARKS

DO NOT
WRITE IN
THIS
MARGIN

9. The graph below shows the average number of units of alcohol consumed per week by men and women in 2006 for an area in the UK.

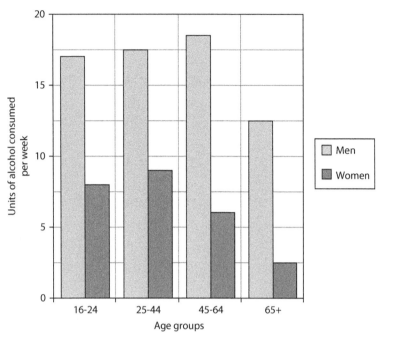

(a) Using information from the graph, calculate the percentage decrease in the number of units of alcohol consumed by a man aged 28 years and a man aged 66 years.

Space for calculation

_____ % **1**

(b) How many times fewer is the number of units of alcohol consumed by a woman aged 66 years compared with a man of a similar age?

Space for calculation

_____ **1**

(c) From the data in the graph, state two different conclusions about the relationship between the age and sex of a person and the average number of units of alcohol consumed in this part of the UK in 2006.

Conclusion 1 _____

_____

_____

Conclusion 2 _____

_____

_____ **2**

MARKS

10. (a) Birds are able to control their internal environment through their metabolism.

   (i) What term is used to describe an animal that can control its internal environment through metabolism?

   _____   **1**

   (ii) Give **one** reason why is it important for birds to control their body temperature.

   _____

   _____

   _____   **1**

(b) State the exact location of the temperature-monitoring centre in mammals.

   _____   **1**

MARKS

DO NOT
WRITE IN
THIS
MARGIN

Practice Papers for SQA Exams: Higher Biology Paper B

11. An experiment was carried out to estimate the sugar content, measured in g/L, of various drinks. The graph below shows the results of the experiment along with the actual sugar content.

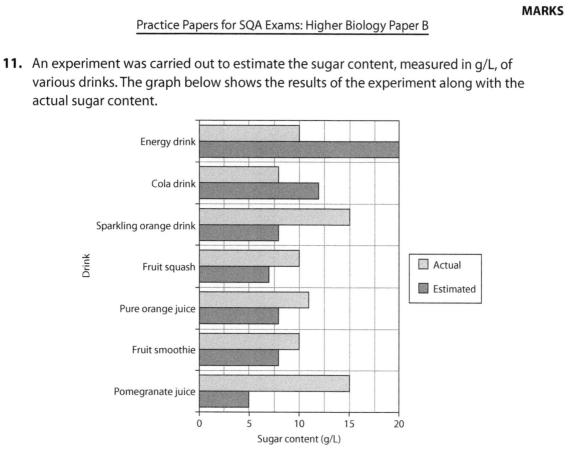

(a) Which drink's estimated sugar content was closest to the actual sugar content?

_____   **1**

(b) Give **one** possible reason why the estimated sugar content was never the same as the actual sugar content.

_____

_____   **1**

MARKS

DO NOT
WRITE IN
THIS
MARGIN

Practice Papers for SQA Exams: Higher Biology Paper B

12. (a) The grid below shows some of the adaptations organisms have in order to survive adverse conditions.

A small ear flap	B hibernation	C thick layer of skin fat
D seed dormancy	E migration	F large body size

Using the **letters**, place each of these adaptations under the correct descriptive heading in the table below.

Structure	*Function*	*Behaviour*

1

(b) State **one** technique used by scientists to monitor long-distance migration in birds and explain how the technique is used.

Technique _____

Explanation _____

_____

_____

2

(c) Migration is affected by both **innate** and **learned** influences.

Give **one** example of each influence.

Innate _____

_____

Learned _____

_____

2

MARKS

DO NOT
WRITE IN
THIS
MARGIN

Practice Papers for SQA Exams: Higher Biology Paper B

13. (a) The diagram below shows some of the reactions of photosynthesis that take place in a chloroplast.

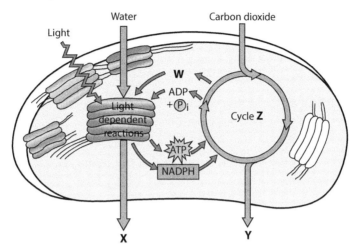

(i) Name molecules **W** and **X**.

W _____

X _____   1

(ii) Name cycle **Z**.

_____   1

(iii) Molecule **Y** can be converted into an important chemical found in the cell wall.

Name this chemical.

_____   1

(b) In eukaryotic cells, molecules involved in photosynthesis are contained within the chloroplast.

Give **one** advantage of this arrangement.

_____

_____

_____   1

(c) A plant was allowed to photosynthesise in normal light. The light was then switched off. There was a rise in the concentration of glyceraldehyde-3-phosphate (G3P).

Give **two** reasons why the concentration of G3P increased.

1 _____

_____

2 _____

_____   2

MARKS

DO NOT
WRITE IN
THIS
MARGIN

Practice Papers for SQA Exams: Higher Biology Paper B

14. In mice, white fur is recessive to black fur.

(a) Using the letter B for the allele controlling black fur colour and b for the allele controlling white fur, what are the possible genotypes for a black-coloured mouse?

_____ 1

(b) State the genotype of a mouse with white fur.

_____ 1

(c) A mouse with white fur can be used to determine the genotype of a black-coloured mouse.

(i) Complete the grid below to show the two possible outcomes of a cross between a mouse with white fur and one with black-coloured fur.

	b	b

	b	b

2

(ii) Name a cross, such as this, which can be used to determine an unknown genotype.

_____ 1

15. Growers often face problems with organisms such as weeds, pests and diseases. To combat these, the grower may use a chemical to protect valuable crop.

(a) What is the general term used to describe such a chemical?

_____ **1**

(b) Some of these chemicals can be **persistent**.

Explain what this term means.

_____

_____ **1**

(c) Give **two** other possible negative effects of using such chemicals.

1 _____

_____

2 _____

_____ **1**

(d) Continued use of such chemicals can lead to **selection pressure**.

State **one** possible outcome of this selection pressure.

_____

_____

_____ **1**

MARKS

DO NOT
WRITE IN
THIS
MARGIN

Practice Papers for SQA Exams: Higher Biology Paper B

16. In an investigation, the rate of photosynthesis by plant leaf discs was measured at different light intensities. The results are shown in the table.

Light intensity (kilolux)	Rate of photosynthesis by leaf discs (units)
5	1
10	20
15	25
20	32
25	45
30	45

(a) Plot a line graph to show the rate of photosynthesis by the plant leaf discs against the light intensity.

2

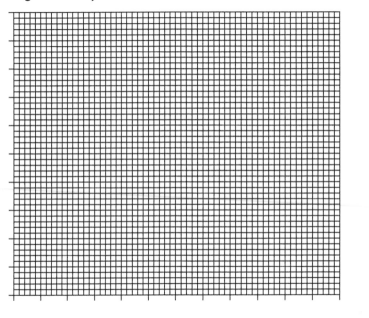

(b) From these results, draw a conclusion about the effect of light intensity on the rate of photosynthesis.

_____

_____

2

(c) Identify the independent variable in this investigation.

_____

1

MARKS

DO NOT
WRITE IN
THIS
MARGIN

Practice Papers for SQA Exams: Higher Biology Paper B

17. Answer **either A or B** in the space below.

A Describe social behaviour in animals under the following headings. 9

   (i) Advantages of social hierarchy

   (ii) Altruism

OR

B Describe genomic sequencing under the following headings. 9

   (i) Molecular clocks

   (ii) Personalised medicine

Labelled diagrams may be used where appropriate.

[END OF QUESTION PAPER]

SECTION 1 ANSWER GRID

Mark the correct answer as shown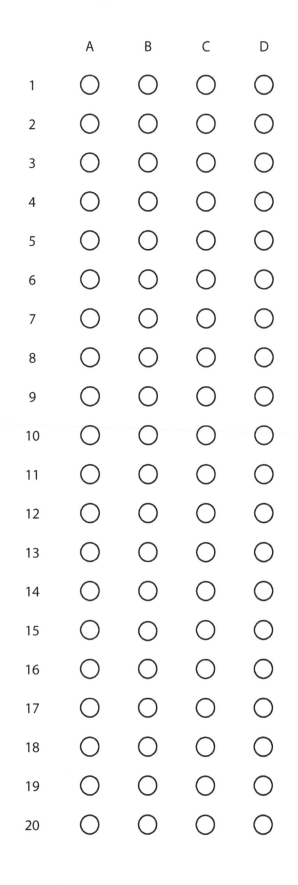

	A	B	C	D
1	○	○	○	○
2	○	○	○	○
3	○	○	○	○
4	○	○	○	○
5	○	○	○	○
6	○	○	○	○
7	○	○	○	○
8	○	○	○	○
9	○	○	○	○
10	○	○	○	○
11	○	○	○	○
12	○	○	○	○
13	○	○	○	○
14	○	○	○	○
15	○	○	○	○
16	○	○	○	○
17	○	○	○	○
18	○	○	○	○
19	○	○	○	○
20	○	○	○	○

Paper C

CfE Higher Biology

Practice Papers for SQA Exams

Paper C

Fill in these boxes and read what is printed below.

Full name of centre

Town

Forename(s)

Surname

Answer all of the questions in the time allowed.

Total marks – 100

Section 1 – 20 marks

Section 2 – 80 marks

Read all questions carefully before attempting.

You have 2 hours 30 minutes to complete this paper.

Write your answers in the spaces provided, including all of your working.

Scotland's leading educational publishers

SECTION 1 – 20 marks

Attempt ALL questions

Answers should be given on the separate answer sheet provided.

1. The graph below shows the temperature changes involved in one cycle of the polymerase chain reaction (PCR).

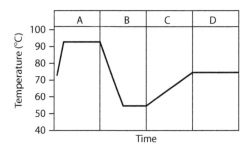

Which letter indicates when primers would bind to target sequences of DNA?

2. Information about DNA, mRNA and tRNA is given below.

Which line in the table is correct?

	DNA	mRNA	tRNA
A	double stranded ribose sugar present	single stranded contains codons	double stranded contains codons
B	single stranded ribose sugar present	single stranded contains anti-codons	single stranded contains anti-codons
C	single stranded deoxyribose sugar present	single stranded contains anti-codons	double stranded contains codons
D	double stranded deoxyribose sugar present	single stranded contains codons	single stranded contains anti-codons

3. Which of the following is a source of stem cells in animals?

A Zygote

B Embryo

C Liver

D Meristem.

4. The genome of an organism is its hereditary information encoded in the

A ribosome

B nuclear membrane

C cytoplasm

D DNA.

5. Which of the following correctly describes the main sequence of events in the evolution of life, with the earliest first?

A Photosynthetic organisms, eukaryotes, multicellular organisms, vertebrates

B Vertebrates, photosynthetic organisms, eukaryotes, multicellular organisms

C Photosynthetic organisms, multicellular organisms, eukaryotes, vertebrates

D Vertebrates, eukaryotes, multicellular organisms, photosynthetic organisms.

6. Potato pieces, of equal size, were weighed then placed in different concentrations of sucrose.

After 24 hours, the potato pieces were removed and reweighed.

For each potato piece, the initial mass divided by the final mass was calculated.

Which graph correctly represents the change in initial mass divided by final mass that would be expected as the concentration of sucrose increases?

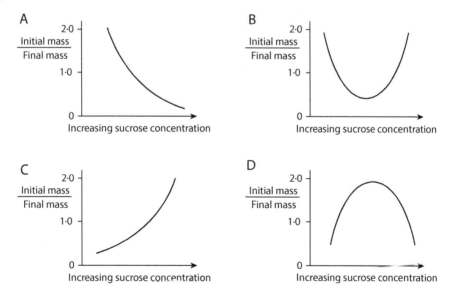

7. The diagram below shows the regeneration of ATP in a cell.

$$ADP + phosphate\ (Pi) \underset{\text{Reaction Y}}{\overset{\text{Reaction X}}{\rightleftarrows}} ATP$$

The following statements relate to this diagram.

1 Reaction Y releases energy for anabolic pathways.

2 Reaction X releases energy for anabolic pathways.

3 Reaction Y is catalysed by ATP synthase.

4 Reaction X is catalysed by ATP synthase.

Which statements are correct?

A 1 and 3

B 2 and 3

C 1 and 4

D 2 and 4

8. In which of the following vertebrate groups can the circulatory system be described as double and incomplete?

A Birds and mammals

B Amphibians and reptiles

C Amphibians and birds

D Mammals and reptiles.

9. Which part of a mammal's body is responsible for monitoring temperature changes and contains central thermoreceptors?

A Hypothalamus

B Heart

C Skin

D Lung.

10. Which of the following correctly describes aestivation?

   A Avoiding metabolic adversity by expending energy to relocate to a more suitable environment

   B A period of reduced activity in organisms with high metabolic rates

   C A form of dormancy employed by some animals to survive periods of excessive heat and drought

   D A form of dormancy that enables some animals to survive the adverse conditions of winter.

11. Which of the following correctly states the culture conditions that must be controlled in the growth of microorganisms?

   A Light intensity, oxygen levels and pH

   B Light intensity, carbon dioxide levels and humidity

   C Temperature, oxygen levels and pH

   D Temperature, carbon dioxide levels and humidity.

12. The following are stages in the process of recombinant DNA technology.

   1 Reprogrammed bacteria identified as having required gene and propagated

   2 Plasmid inserted into host bacterial cells

   3 DNA extracted and cut into many fragments using endonuclease

   4 DNA fragments sealed into plasmids using ligase.

Which sequence below, starting with the earliest, places the above steps into the correct order?

   A $3 \rightarrow 4 \rightarrow 2 \rightarrow 1$

   B $4 \rightarrow 3 \rightarrow 1 \rightarrow 2$

   C $3 \rightarrow 4 \rightarrow 1 \rightarrow 2$

   D $4 \rightarrow 1 \rightarrow 3 \rightarrow 2$

13. The graph below shows the net energy gain or loss from hunting and eating prey of different masses.

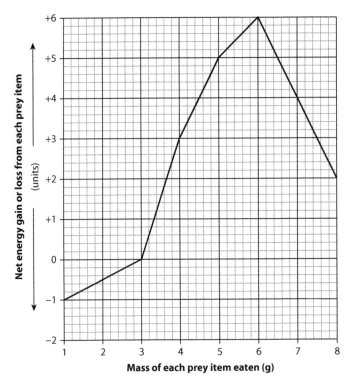

It can be concluded from the graph that

A prey weighing between 1 g and 3 g are rarer than prey weighing between 3 g and 6 g

B hunting and eating prey weighing above 6 g involves a net energy loss

C prey weighing 8 g contain less energy than prey of mass 6 g

D hunting and eating prey weighing below 3 g involves a net energy loss.

14. The graph below shows the annual variation in the biomass and population density of *Bourletiella viridescens,* found primarily in the Cairngorms in the UK.

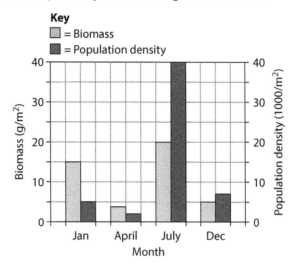

During which month do individual *Bourletiella viridescens* have the greatest average mass?

A January

B April

C July

D December.

15. Which of the following are properties of annual weeds?

A Steady growth, short life cycle, low seed output, short-term seed viability

B Rapid growth, long life cycle, low seed output, long-term seed viability

C Steady growth, long life cycle, high seed output, short-term seed viability

D Rapid growth, short life cycle, high seed output, long-term seed viability.

16. Which behavioural indicator may involve an animal mutilating itself by chewing on its own hair?

A Stereotypical

B Misdirected

C Appeasement

D Failure in parental behaviour.

17. Which definition best describes mutualism?

A One organism benefits at the expense of the other organism, illustrating dependence

B Neither organism benefits from the relationship

C One organism benefits at the expense of the other organism, illustrating interdependence

D Both organisms benefit from the relationship, illustrating interdependence.

18. Which of the following forms of biodiversity are measurable?

A Genetic, species, ecosystem

B Genetic, species, competition

C Species, ecosystem, competition

D Competition, genetic, ecosystem.

19. An experiment was carried out to determine the effect of substrate concentration on enzyme activity, as shown in the diagram below. The method involved adding pieces of fresh liver to solutions of hydrogen peroxide, each at a different concentration. The height of the froth produced from the reaction was measured using a ruler.

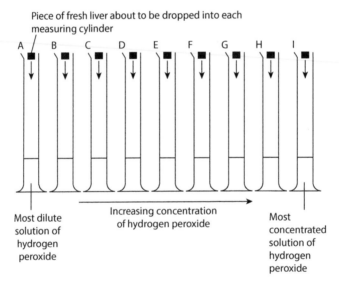

Which two factors would have to be kept the same throughout the investigation?

A Size of liver piece and concentration of hydrogen peroxide

B Concentration of hydrogen peroxide and height of froth measured

C Size of liver piece and volume of hydrogen peroxide used

D Volume of hydrogen peroxide used and height of froth measured.

MARKS

DO NOT
WRITE IN
THIS
MARGIN

Practice Papers for SQA Exams: Higher Biology Paper C

20. The graph below shows changes in the number of breeding pairs of peregrine falcons between 1950 and 1990.

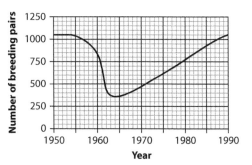

Which line in the table correctly shows the overall increase in the 10-year periods shown?

	10-year period	**Increase in number of breeding pairs**
A	1950 and 1960	250
B	1960 and 1970	350
C	1970 and 1980	400
D	1980 and 1990	275

MARKS

SECTION 2 – 80 marks

Attempt ALL questions

It should be noted that questions 5 and 13 contain a choice.

1. The diagram shows a strand of DNA during replication and associated structures.

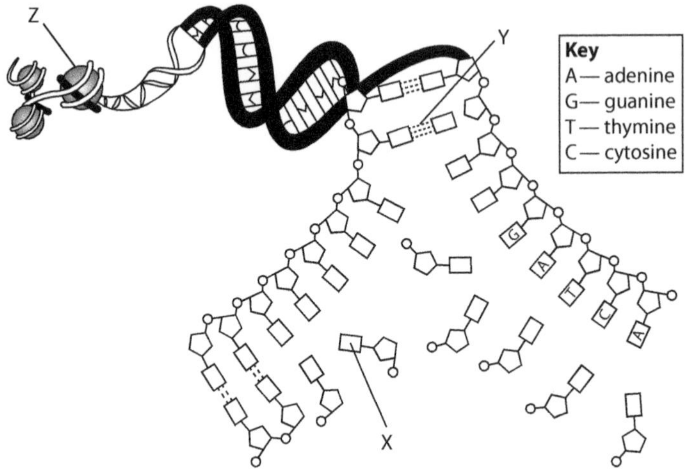

Key
A— adenine
G— guanine
T— thymine
C— cytosine

(a) Name base **X**.

_____ **1**

(b) Name bond **Y**.

_____ **1**

(c) State the function of molecule **Z**.

_____

_____ **1**

(d) Explain the function of a primer.

_____

_____ **1**

(e) State the name of the enzyme that joins fragments of DNA together.

_____ **1**

(f) The polymerase chain reaction (PCR) is a technique that can be used to amplify DNA.

Describe the main steps involved in this technique.

_____

_____

_____

_____ **3**

MARKS

2. (a) Decide if each of the statements relating to genomic sequencing in the table below is true or false and tick (✓) the appropriate box.

If you decide that the statement is false, write the correct term in the correction box to replace the term <u>underlined</u> in the statement.

Statement	True	False	Correction
<u>Phylogenetics</u> is the study of evolutionary relatedness.			
The study of comparative gene sequences using computers and statistics is called <u>pharmacogenetics</u>.			
<u>Sequence</u> data is used to study the evolutionary relatedness among groups of organisms.			

3

(b) The diagram below shows a phylogenetic tree.

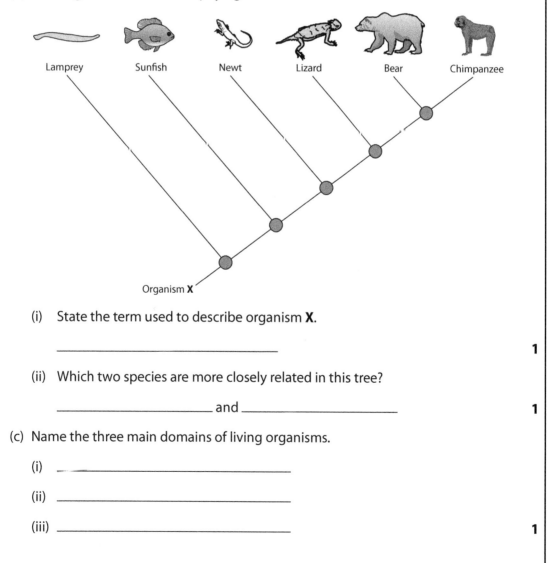

(i) State the term used to describe organism **X**.

_____

1

(ii) Which two species are more closely related in this tree?

_____ and _____

1

(c) Name the three main domains of living organisms.

(i) _____

(ii) _____

(iii) _____

1

3. Single gene mutations involve a change in one of the base pairs in the DNA sequence of a gene.

(a) State where a single gene mutation may occur.

_____ **1**

(b) Suggest one possible effect of a single gene mutation on gene expression.

_____ **1**

Below are three different types of single gene mutation.

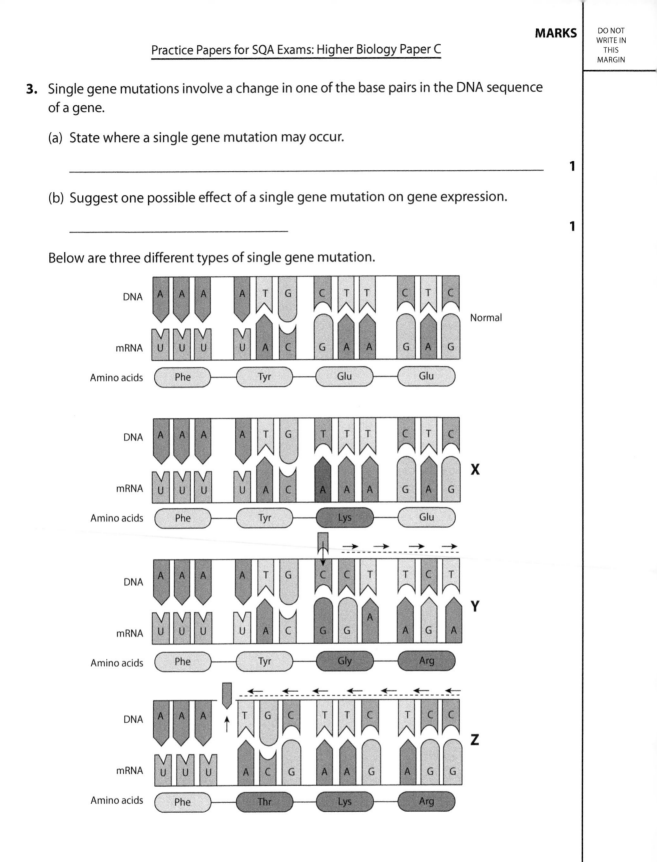

MARKS

DO NOT
WRITE IN
THIS
MARGIN

Practice Papers for SQA Exams: Higher Biology Paper C

(c) Name each type of single gene mutation.

X _____

Y _____

Z _____ **3**

(d) Explain the evolutionary importance of mutation.

_____

_____

_____ **2**

4. Two different species of *Paramecia* were grown – *Paramecium aurelia* and *Paramecium caudatum* – and the population growth is displayed in the graph below.

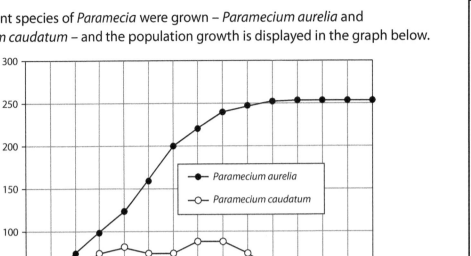

(a) Describe the change in *Paramecium caudatum* population between days 6 and 14.

_____

_____

_____ **2**

(b) Calculate the percentage change in the *Paramecium aurelia* population between days 3 and 10.

Space for calculation

_____ % **1**

MARKS

(c) Both species of *Paramecia* were grown under specific culture conditions.

State two conditions that must be controlled when growing microbial cultures.

1 _____

2 _____   **2**

(d) Microorganisms require media to stimulate growth.

Give an example of a complex substance that may be added to media.

_____   **1**

(e) Another culture experiment was set up, using the same conditions for the two species of *Paramecium*. The results are shown in the table below.

Time (days)	Number of colonies of microorganism X
0	0
2	1
4	3
6	8
8	18
10	44

On the grid below, plot a line graph to show the population growth of the microorganism.   **2**

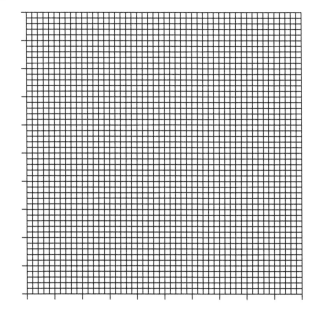

MARKS

5. Answer **either A or B** in the space below.

   A Describe protein synthesis under the following headings.   **8**

      (i)   transcription

      (ii) translation.

   OR

   B Describe social behaviour under the following headings.   **8**

      (i)   kin selection

      (ii) primate behaviour.

   Labelled diagrams may be used where appropriate.

MARKS

DO NOT
WRITE IN
THIS
MARGIN

Practice Papers for SQA Exams: Higher Biology Paper C

6. The diagram below shows the first two stages of cellular respiration.

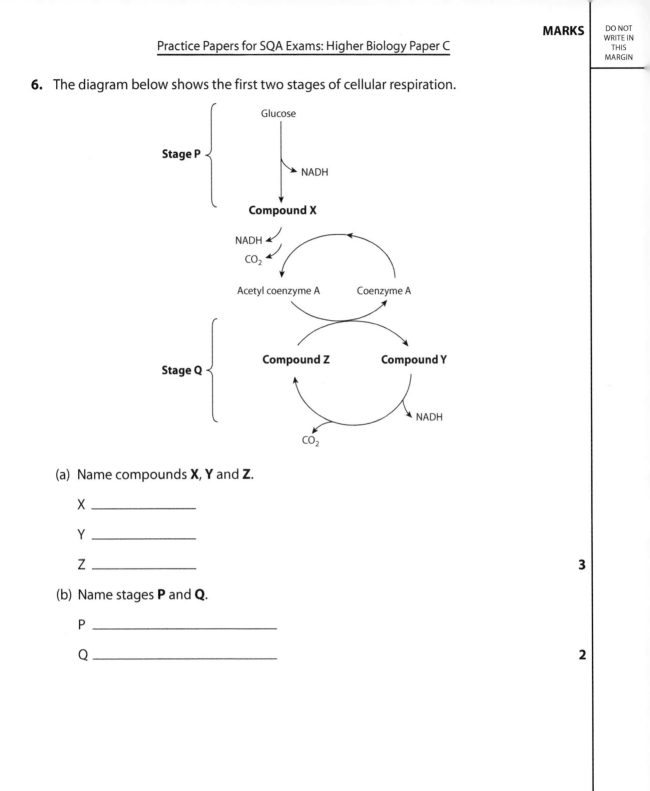

(a) Name compounds **X**, **Y** and **Z**.

X _____

Y _____

Z _____

3

(b) Name stages **P** and **Q**.

P _____

Q _____

2

MARKS

DO NOT
WRITE IN
THIS
MARGIN

Practice Papers for SQA Exams: Higher Biology Paper C

(c) State the final destination of the hydrogen in NADH, produced in stages **P** and **Q**.

_____   **1**

Below is a diagram summarising the breakdown of two respiratory substrates,
L and **M**, and their role in cellular respiration.

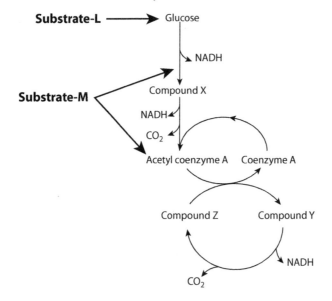

(d) Name respiratory substrates **L** and **M**.

L _____

M _____   **2**

(e) State another respiratory substrate, not shown on this diagram, which is used
only during prolonged starvation.

_____   **1**

7. The apparatus shown below was used to measure the metabolic rate of a human subject at rest.

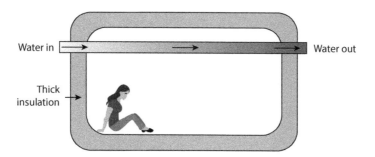

Water in →   →   → Water out

Thick insulation →

(a) Give the name of this apparatus.

_____   **1**

(b) Describe how this chamber measures metabolic rate.

_____

_____   **2**

(c) Another way to measure metabolic rate is to measure oxygen consumption over a period of time. The higher the oxygen consumption, the higher the metabolic rate. The table below shows the metabolic rates of different organisms at rest.

Organism	Weight (kg)	Oxygen consumption (cm³/min)
A	12	110
B	34	306
C	62	405
D	19	150
E	84	490

What conclusion can be drawn, from the data above, about the relationship between the weight of the organism and oxygen consumption?

_____

_____   **1**

MARKS

(d) Calculate the oxygen consumption of organism B per kg of its body weight.

Space for calculation

_____ cm³/kg/min   **1**

(e) Select the organism, from the list shown, which consumes the most oxygen per kg of body weight, per minute.

Space for calculation

Organism _____   **1**

MARKS

DO NOT
WRITE IN
THIS
MARGIN

Practice Papers for SQA Exams: Higher Biology Paper C

8. Some bacteria live in hot springs at temperatures such as 50–80°C.

(a) State the name used to describe organisms like the bacteria above.

_____ **1**

(b) Give an example of an enzyme from these bacteria that has been used to amplify DNA and give a use for this procedure.

Enzyme _____

Use _____

_____ **2**

9. A field trial, investigating the effect of different soil types upon potato yield, was set up as shown below.

Soil type 1 (clay)	Soil type 2 (sandy)
Soil type 4 (silty)	Soil type 3 (peaty)

(a) State the independent variable in this investigation.

_____ **1**

(b) Suggest a reason why the four plots are placed side-by-side, instead of being separated by a larger distance.

_____

_____ **1**

(c) Describe **two** other considerations to be taken into account when designing a field trial such as the one above.

_____

_____ **2**

(d) The results of the investigation are shown below.

Soil type	Seasonal yield (kg)
1	40·8
2	0·0
3	204·0
4	79·2

Express, as the simplest whole number ratio, the seasonal yield of soil type 1 to that of soil type 3.

Space for calculation

_____ : _____ **1**

10. The graph below shows the percentage of living species becoming extinct over the last 300 million years.

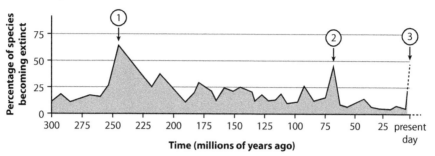

(a) State the name given to events **1** and **2**.

_____ 1

(b) Describe the changes in extinction rate between 250 million and 200 million years ago.

_____

_____ 2

(c) Event **3** suggests that a high number of species could become extinct in the near future. Suggest a possible cause of this.

_____

_____ 1

(d) From the graph, describe the change in biodiversity immediately after events **1** and **2**.

_____ 1

11. Four different varieties of crop were planted in four adjacent fields, as shown below.

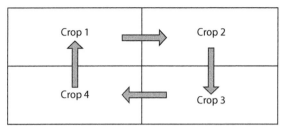

Each year, the crops are planted in a different field as shown by the arrows.

(a) State the term used for this process.

_____   **1**

(b) As well as the method above, describe two other ways in which crops are protected by cultural means.

1 _____

_____

2 _____

_____   **2**

(c) Crop 1 is infested with a known pest. A natural predator to this pest is introduced into this field.

State the term used to describe the predator in this instance.

_____   **1**

(d) Describe **one** risk of using the above method.

_____

_____   **1**

12. An experiment was set up to determine whether dogs were more motivated by food or access to a sexually receptive mate.

Each dog had to choose whether to enter a room containing only food, or a room with only a sexually receptive mate.

Four groups of ten dogs were selected, with each group comprising a range of dog breeds, five male and five female. All dogs were aged 3 years.

In the week prior to the experiment, each group had a controlled level of access to food and mates, shown in the table below.

Group	Meals a day	Access to a mate
1	3	Yes
2	3	No
3	1	Yes
4	1	No

The results from the experiment are shown below.

	Choice (number of dogs)	
Group	Food	Mate
1	9	1
2	4	6
3	8	2
4	4	6

(a) State the term used to describe this type of choice experiment.

_____ **1**

(b) Calculate the percentage of dogs that chose to enter the room with food.

Space for calculation

_____ % **1**

(c) Express, as the simplest whole number ratio, the proportion of dogs choosing food to those choosing mates.

Space for calculation

MARKS

(d) Give **one** conclusion that can be drawn when comparing the results from group 1 and 2.

_____

_____

_____

_____   **2**

(e) Suggest **one** improvement that could be made to this experiment to improve the reliability of the results.

_____   **1**

13. Answer **either A or B** in the space below.

    A Write notes on the ethical considerations, risks and hazards in the use of microorganisms.   **4**

    OR

    B Write notes on the ethical issues of stem cell use.   **4**

    Labelled diagrams may be used where appropriate.

[END OF QUESTION PAPER]

SECTION 1 ANSWER GRID

Mark the correct answer as shown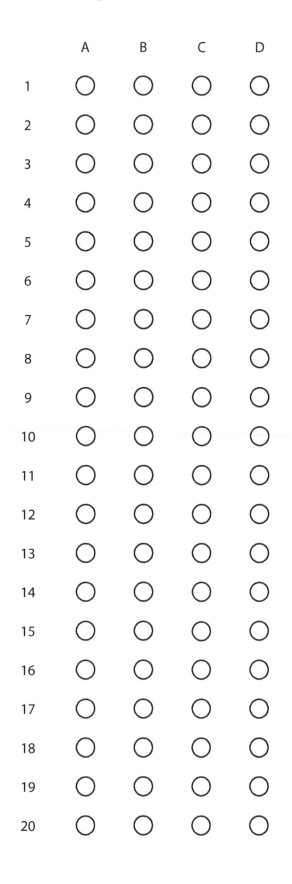

	A	B	C	D
1	○	○	○	○
2	○	○	○	○
3	○	○	○	○
4	○	○	○	○
5	○	○	○	○
6	○	○	○	○
7	○	○	○	○
8	○	○	○	○
9	○	○	○	○
10	○	○	○	○
11	○	○	○	○
12	○	○	○	○
13	○	○	○	○
14	○	○	○	○
15	○	○	○	○
16	○	○	○	○
17	○	○	○	○
18	○	○	○	○
19	○	○	○	○
20	○	○	○	○

Paper D

CfE Higher Biology

Practice Papers for SQA Exams

Paper D

Fill in these boxes and read what is printed below.

Full name of centre

Town

Forename(s)

Surname

Answer all of the questions in the time allowed.

Total marks – 100

Section 1 – 20 marks

Section 2 – 80 marks

Read all questions carefully before attempting.

You have 2 hours 30 minutes to complete this paper.

Write your answers in the spaces provided, including all of your working.

Scotland's leading educational publishers

SECTION 1 – 20 marks

Attempt ALL questions

Answers should be given on the separate answer sheet provided.

1. The following graph shows how the heartbeat of a species of a particular fish varies with the length of development at three different temperatures.

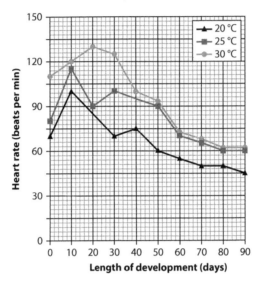

Which of the following is a valid conclusion from this data?

A At all temperatures, the heart rate increases with increasing length of development

B After 50 days, the heart rate at 25 °C is 30 beats per minute faster than at 20 °C

C The greatest change in heart rate over the 90 days from the start was at 20 °C

D The lowest heart rate was recorded at 25 °C.

2. Which of the following would **not** describe post-translational modification of a protein?

A Addition of a phosphate group to make a protein function

B The process does not require the presence of enzymes

C Addition of a sugar to a carbohydrate group

D Usually occurs after the protein has folded and coiled.

3. What feature of stem cells makes them useful in medicine?
They can develop into

    A nerve cells for stroke victim patients

    B organs for transplantation

    C reproductive tissues

    D many different types of tissue.

4. A missense point mutation

    A is a substitution and the protein formed may function in a different way

    B brings the translation process to a complete halt

    C causes all the subsequent codons to be altered

    D does not affect the amino acid coded for by the altered codon.

5. Recently, a strain of the bacterium *Staphylococcus aureus* has emerged which is highly resistant to a number of antibiotics. Predict what would happen if the use of these antibiotics was stopped.

    A The resistant strain would disappear entirely

    B Populations of the resistant strain would colonise new environments

    C Populations of the non-resistant variety of this strain would increase

    D Resistant forms would increase markedly.

6. The study of how groups of organisms are related through evolution is called

    A comparative genomics

    B phylogenetics

    C gene sequencing

    D bioinformatics.

7. The function of the high-energy electrons in the electron transport chain is to

    A combine with phosphate to form ATP

    B react with ATP synthase

    C pump hydrogen ions against a concentration gradient

    D pick up energy as they move along the chain.

8. Which of the following is **not** a suitable method for measuring the metabolic rate of an organism?

   A Heat produced in unit time

   B Oxygen consumed in unit time

   C Carbon dioxide produced in unit time

   D Heart rate in unit time.

9. Conformers

   A are able to regulate their metabolism and maintain a steady internal environment

   B use physiological mechanisms to alter their metabolic rate

   C depend on behaviour to regulate their internal state

   D live in unstable environments.

10. A short-term physiological state in which the metabolic rate and body temperature of mammals are both lowered at night is called

    A hibernation

    B aestivation

    C torpor

    D adaptation.

11. Extremophiles are most commonly found in

    A algae

    B protozoa

    C moulds

    D bacteria.

12. The following are stages in the production of a human hormone using DNA recombinant technology.

1 Production of hormone by transformed bacteria

2 DNA sealed into bacterial plasmid

3 DNA extracted from normal human chromosome using restriction endonuclease

4 Required genes on human chromosome identified

What is the correct order of these stages, starting with the earliest?

A $1 \rightarrow 2 \rightarrow 3 \rightarrow 4$

B $3 \rightarrow 4 \rightarrow 2 \rightarrow 1$

C $2 \rightarrow 4 \rightarrow 1 \rightarrow 3$

D $4 \rightarrow 3 \rightarrow 2 \rightarrow 1$

13. Which of the following could lead to obtaining inaccurate results from an experiment?

A Recording the behaviour of an animal from a concealed observation tower

B Inserting a cold thermometer into warm water to measure the temperature

C Repeating an experiment many times

D Using a digital balance to measure a change in mass.

14. A microorganism that has been genetically modified may have an increased ability to cause disease in humans.

The following are some of the stages used in assessing the risk involved in this process.

1 Set up suitable control measures

2 Identify the hazard

3 Evaluate the effectiveness of the control measures.

What is the correct order of these stages, starting with the first stage?

A $1 \rightarrow 2 \rightarrow 3$

B $1 \rightarrow 3 \rightarrow 2$

C $2 \rightarrow 3 \rightarrow 1$

D $2 \rightarrow 1 \rightarrow 3$

15. In which of the following parts of the absorption spectrum of chlorophyll is absorption at the maximum?

A Blue

B Green

C Red

D Yellow.

16. In a genetics experiment involving plants that produced different coloured flowers, 100 flowers were produced, of which 50 were white, 40% were red and the rest blue.

Which of the following correctly describes the ratio of blue to red to white flowers?

A 1:5:4

B 5:1:4

C 4:5:1

D 1:4:5

17. A student carried out an experiment to look at the growth of a plant under different conditions over a 24-hour period. The results are summarised in the table below.

Conditions within a 24-hour period		
Length of exposure to light (hours)	Volume of water added (cm³)	Growth (cm)
10	0	0
10	2	0·5
10	4	1·0
10	6	1·2
10	8	1·9

Which of the following is correct?

	Controlled variable	Independent variable	Dependent variable
A	length of light exposure	volume of water added	growth measured
B	growth measured	volume of water added	length of light exposure
C	volume of water added	growth measured	length of light exposure
D	length of light exposure	growth measured	volume of water added

18. Inbreeding can lead to

    A an increase in the genotype that will produce a dominant phenotype

    B a decrease in the genotype that will produce a recessive phenotype

    C an increase in the frequency of homozygosity

    D an increase in genetic heterozygosity.

19. An advantage to an animal living in a social group is

    A increased competition for resources

    B less energy is wasted

    C reproduction rates are lowered

    D less cooperation in caring for young.

20. The graph below shows changes in the mass of protein, fat and carbohydrate in an animal's body during a period of food reduction that lasted for 5 weeks.

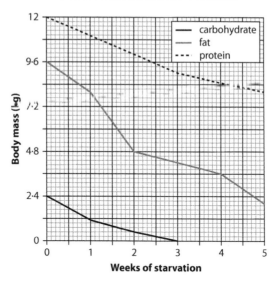

The animal's initial mass was 50 kg. From this data alone, predict the animal's weight after 4 weeks.

    A 48 kg

    B 35 kg

    C 38 kg

    D 30 kg.

MARKS

SECTION 2 – 80 marks

Attempt ALL questions

It should be noted that questions 5 and 16 contain a choice.

1. (a) Complete the following table showing some of the differences that exist between bacterial cells and animal cells by writing appropriate answers into the blank spaces in the table.

2

Feature	Bacterial cell	Animal cell
ribosomes	present	
mitochondria		present
plasmids	often present	
DNA		arranged into linear chromosomes

(b) The diagram below shows the arrangement of part of a DNA molecule found in animal cells.

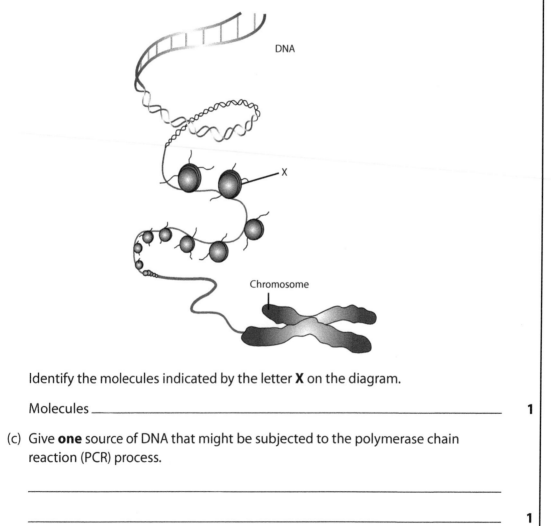

DNA

X

Chromosome

Identify the molecules indicated by the letter **X** on the diagram.

Molecules _____ **1**

(c) Give **one** source of DNA that might be subjected to the polymerase chain reaction (PCR) process.

_____

_____ **1**

2. A small farm pond was accidentally polluted with a fertiliser. The changing numbers of two different species of a pond invertebrate – *Daphnia magna* and *Daphnia pulex* – were measured over a period of 9 days following the pollution event. The results are shown in the table below.

| Day | Numbers | |
	Daphnia magna (per l)	*Daphnia pulex* (per l)
1	20	10
2	30	15
3	120	30
4	130	55
5	170	60
6	180	65
7	182	70
8	185	90
9	185	120

(a) On the grid below, plot **two** line graphs to show the change in numbers of each species against the number of days after the pollution event.

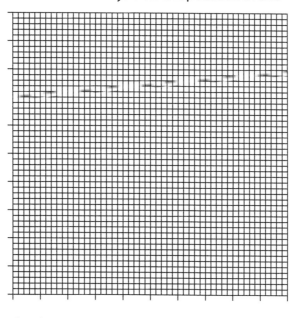

2

(b) From these results, draw **two** conclusions about the effect of the pollution on the numbers of the two species of *Daphnia*.

1 _____

_____

_____

_____

1

MARKS

2 _____

_____

_____

_____ **1**

(c) Calculate the percentage increase in the numbers of *Daphnia pulex* from day 1 to day 5.

Space for calculation

_____% **1**

MARKS

DO NOT
WRITE IN
THIS
MARGIN

Practice Papers for SQA Exams: Higher Biology Paper D

3. A company that grows plants commercially uses a technique called hydroponics, which does not involve soil.
The plants are grown in reservoirs of nutrient solution as shown.

Nutrient solution in reservoir

The nutrient solution contains water and other elements the plants need such as nitrogen and phosphorus. The company has been told that adding iron to the nutrient solution will greatly enhance the growth of the plants and so they decide to test this hypothesis.

(a) Suggest **one** way in which the growth of the plants with or without iron being added to the nutrient solution could be measured.

_____

_____ 1

(b) (i) State the independent variable in this investigation.

Independent variable _____

_____ 1

(ii) State **two** variables, not already mentioned, which would have to be kept constant.

1 _____

2 _____ 1

(c) How would a control group differ from the experimental group?

_____

_____

_____ 1

MARKS

4. (a) The diagram below shows some of the events of protein synthesis.

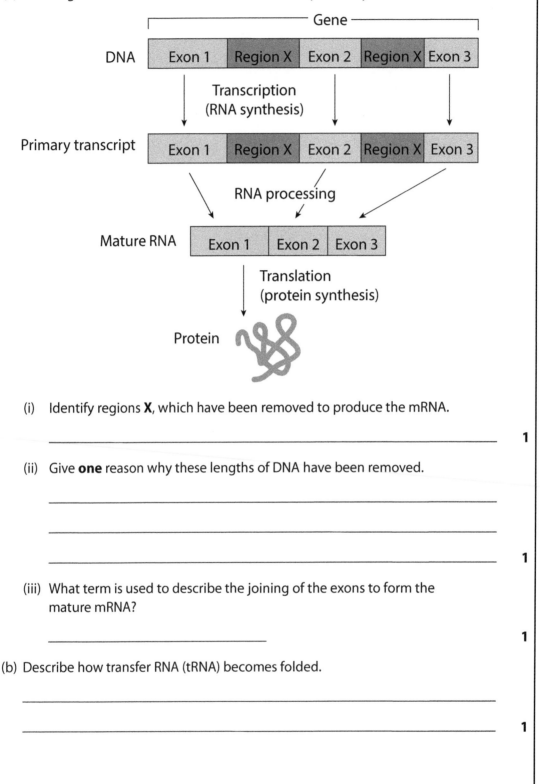

(i) Identify regions **X**, which have been removed to produce the mRNA.

_____ 1

(ii) Give **one** reason why these lengths of DNA have been removed.

_____

_____

_____ 1

(iii) What term is used to describe the joining of the exons to form the mature mRNA?

_____ 1

(b) Describe how transfer RNA (tRNA) becomes folded.

_____

_____ 1

MARKS

DO NOT
WRITE IN
THIS
MARGIN

Practice Papers for SQA Exams: Higher Biology Paper D

5. Answer **either A or B.**

 A Describe the therapeutic uses of stem cells. **4**

 OR

 B Describe how metabolism can be regulated by competitive inhibitors. **4**

 Labelled diagrams may be used where appropriate.

MARKS

6. (a) The following diagram shows a possible food chain.

sun → grass → locust → lizard → snake → eagle

Only about 10% of the energy from one level is incorporated into the tissues of the next organism in line.

Give **two** reasons why energy is lost between each feeding level.

1 _____

_____

2 _____

_____ **2**

(b) Give **one** reason why short food chains are more efficient than long food chains.

_____

_____

_____ **1**

7. An experiment was carried out on a species of fruit fly to determine its response to gravity and different wavelengths of light. Twenty flies were put into each of four different glass containers, which were then sealed and numbered 1–4. Tubes 1 and 2 were half-covered with light-proof paper.

(a) In one experiment, all four tubes were illuminated equally from above by blue light for 10 minutes, orientated as shown.

The number of flies in each clear part after 10 minutes is also shown.

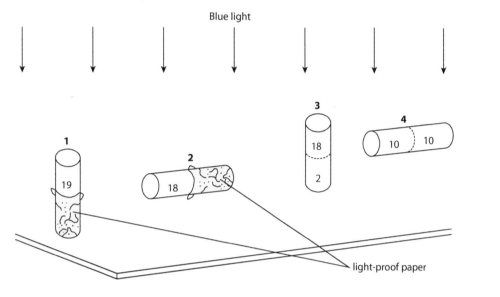

With respect to the blue light and gravity, what was the response of the fruit flies to these two variables?

Blue light _____

_____ **1**

Gravity _____

_____ **1**

(b) Which tube(s) test(s) the response of the flies to gravity alone?

_____ **1**

(c) How could the results of this experiment be made more reliable?

_____ **1**

MARKS

(d) The diagram **A** below shows a clinostat, a device that can slowly rotate. If one of the experimental tubes containing 20 flies was fixed to the disc as shown in diagram **B** and rotated for 10 minutes under blue light, predict the distribution of the flies and give an explanation for you answer.

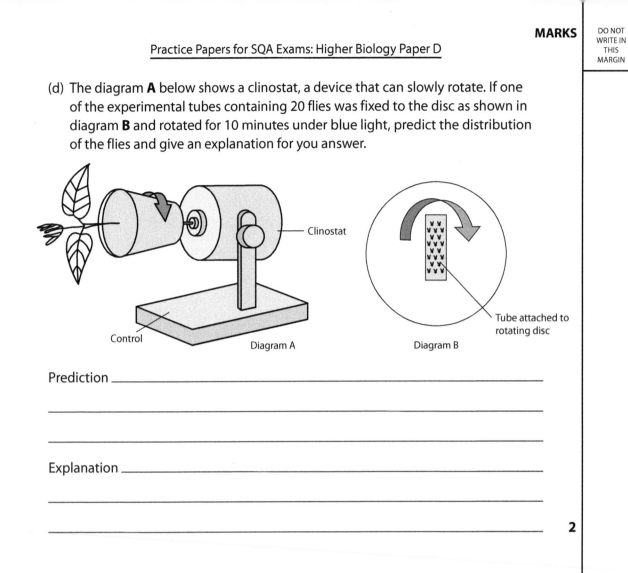

Clinostat

Control

Diagram A

Tube attached to rotating disc

Diagram B

Prediction _____

_____

_____

Explanation _____

_____

_____

2

MARKS

8. The diagram below shows the normal sequence of amino acids coded for by a segment of template DNA as well as some different types of point mutations. The amino acids being coded for are represented by their conventional abbreviations.

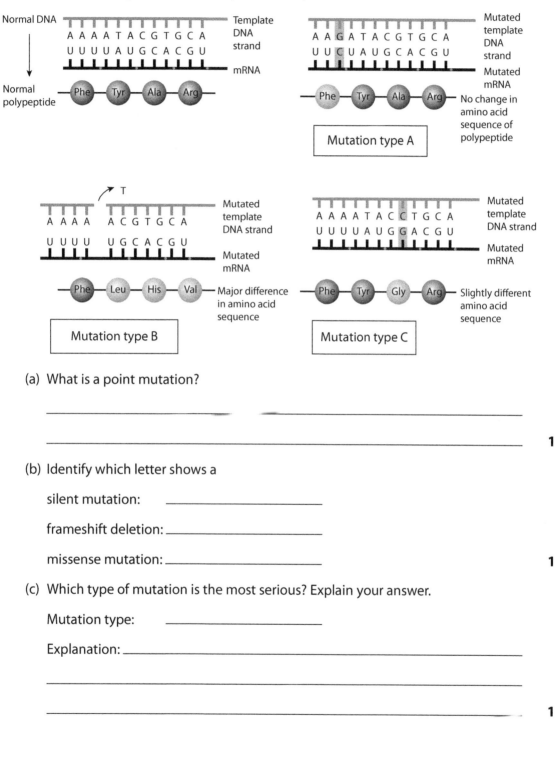

(a) What is a point mutation?

_____

_____   **1**

(b) Identify which letter shows a

silent mutation: _____

frameshift deletion: _____

missense mutation: _____   **1**

(c) Which type of mutation is the most serious? Explain your answer.

Mutation type: _____

Explanation: _____

_____

_____   **1**

9. The diagram below shows the puffer fish *Fugu rubripes*.

Among vertebrates, this fish has the shortest genome, much smaller than that of a human, but the number of genes is much higher than that of humans.

(a) What feature of the DNA found in a puffer fish accounts for the much smaller size of its genome compared with the human genome?

_____

_____

_____

_____ **2**

(b) Give **one** reason why this fish is of use to scientists working on genomic sequencing.

_____

_____

_____ **1**

MARKS

10. The two diagrams below show different types of metabolic pathways.

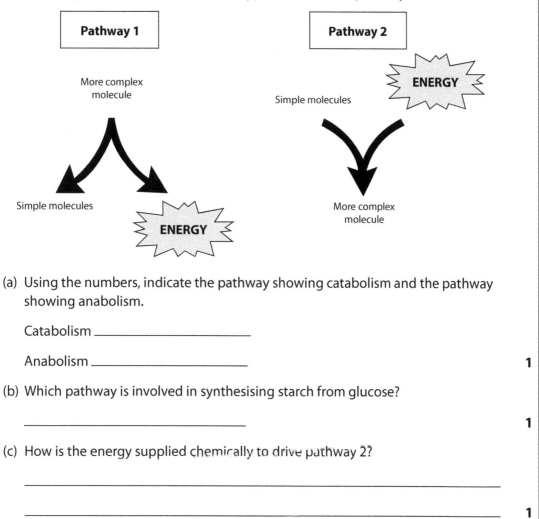

(a) Using the numbers, indicate the pathway showing catabolism and the pathway showing anabolism.

Catabolism _____

Anabolism _____   **1**

(b) Which pathway is involved in synthesising starch from glucose?

_____   **1**

(c) How is the energy supplied chemically to drive pathway 2?

_____

_____   **1**

11. The diagram below shows two important cellular organelles, their compartments and membranes.

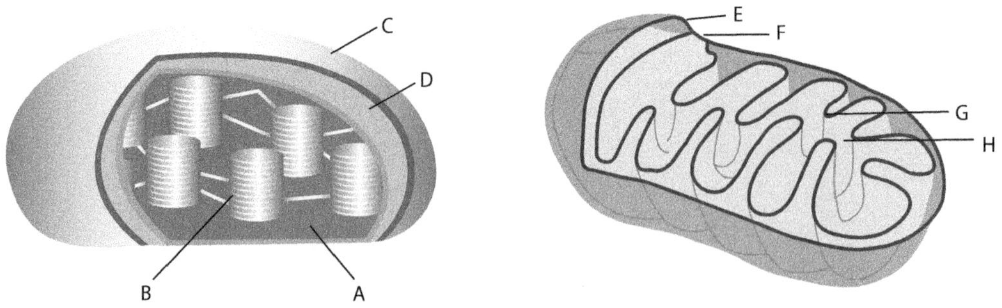

| **Organelle X** | | **Organelle Y** |

(a) Identify these two organelles.

Organelle X: _____

Organelle Y: _____ **1**

(b) Using the letters, indicate where each of the following would be found:

(i) enzymes associated with the citric acid cycle _____

(ii) enzymes associated with the Calvin cycle _____ **1**

(c) Give **two** advantages of inner membranes of organelles being highly folded.

1 _____

_____

2 _____

_____ **2**

12. (a) On Earth, sunlight energy is approximately $2 \cdot 0 \times 10^{12}$ kilojoules per 10 000 square metres. Energy provided by photosynthesis is approximately $2 \cdot 0 \times 10^{10}$ kilojoules per 10 000 square metres.

Calculate the percentage efficiency of photosynthesis.

Space for calculation

_____ % **1**

(b) To produce 1 unit of glucose, the plants referred to in the table below need to receive the following:

2 units of water

2 units of carbon dioxide

4 units of light energy

Plant	Units of carbon dioxide available to plant	Units of water available to plant	Units of light energy available to plant
1	4	4	7
2	6	8	16
3	4	8	4
4	6	8	11

In which of the plants was photosynthesis limited by the availability of carbon dioxide?

Explain your answer.

Plant _____ **1**

Explanation _____

_____

_____ **1**

MARKS

13. (a) The enzyme luciferase is found in organisms that produce light, such as fireflies.

Luciferase catalyses the breakdown of luciferin using ATP as an energy source to produce light.

$$\text{luciferin} + \text{ATP} \xrightarrow{\quad\text{luciferase}\quad} \text{end products} + \text{light energy}$$

(i) ATP is used up in this reaction. What term is used to describe the addition of phosphate to ADP to generate more ATP?

_____   **1**

(ii) Assuming that luciferin and luciferase are in plentiful supply, what will limit the light energy produced?

_____   **1**

(iii) This reaction is about 80% efficient in converting the energy contained in ATP to light energy.

Explain how this reaction is potentially a good way of determining the unknown ATP concentration in a solution.

_____

_____

_____

_____   **2**

MARKS

DO NOT
WRITE IN
THIS
MARGIN

Practice Papers for SQA Exams: Higher Biology Paper D

(b) The diagram below shows part of the mechanism for generating ATP found in the mitochondrion.

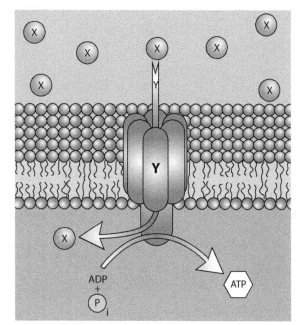

(i) What is represented by each of letters indicated on the diagram?

X _____

Y _____   **2**

(ii) What effect does the flow of **X** have on structure **Y**?

_____

_____   **1**

(c) Organisms do not need to have large stores of ATP.

Give **one** reason for this.

_____

_____

_____   **1**

MARKS

14. (a) The photograph below shows an athlete having his VO_2 max measured.

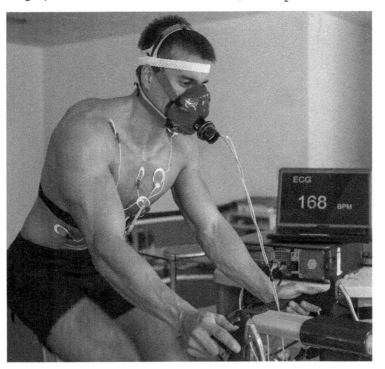

    (i) What is VO_2 max?

_____

_____

_____

_____ **2**

    (ii) Give **one** reason why an athlete might want to have his/her VO_2 max measured.

_____

_____ **1**

    (iii) What happens to the VO_2 max in response to each of the following?

Increased levels of training _____

Increased age _____ **1**

(b) Give **one** example of how mammals such as dolphins and seals are adapted to make extended deep dives.

_____

_____ **1**

MARKS

DO NOT
WRITE IN
THIS
MARGIN

Practice Papers for SQA Exams: Higher Biology Paper D

15. The diagram below shows the appearance of a plate containing a medium that enhances the growth of bacteria that can fix nitrogen. This medium had all the essential nutrients for bacterial growth but no nitrogen-containing ammonium compounds. The plate is shown after 3, 7 and 12 days of growth.

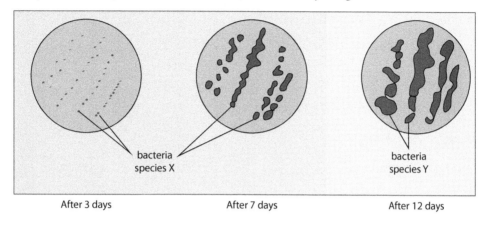

| | | |
| After 3 days | After 7 days | After 12 days |

(a) After 3 days only, species **X** was seen to grow. Suggest why this happened.

_____

_____

_____   **2**

(b) After 12 days, species **Y** appears. Give a possible explanation for this.

_____

_____

_____   **2**

MARKS

16. Answer either **A** or **B** in the space below.

A Describe the factors which influence agricultural productivity under the following headings. **9**

(i) Plant productivity

(ii) Animal productivity

OR

B Describe mass extinction under the following headings. **9**

(i) Mass extinction events

(ii) Measuring extinction rate

Labelled diagrams may be used where appropriate.

MARKS

DO NOT
WRITE IN
THIS
MARGIN

Practice Papers for SQA Exams: Higher Biology Paper D

17. (a) (i) State **two** ways in which the physical well-being of animals used by humans may be judged.

1 _____

_____

2 _____

_____ **1**

(ii) Give **two** examples of behavioural indicators of poor animal welfare.

1 _____

_____

2 _____

_____ **1**

(b) What is a **preference test** and how can it be used in studying animal welfare?

Preference test _____

_____

_____

How used _____

_____

_____

_____ **2**

(c) In the study of animal welfare, a catalogue of behaviours or displays by animals may be constructed.

What is the name of such a catalogue? _____ **1**

18. (a) Give the terms that describe the symbiotic relationship between two organisms when

(i) one organism benefits while the other may be damaged

_____

(ii) both organisms derive benefit

_____ **1**

(b) What is meant by a disease vector?

_____

_____ **1**

(c) It has been suggested that some cell organelles may have their origins among bacteria.

Give **one** example of such an organelle and explain how this may have happened.

Example: _____

Explanation: _____

_____

_____ **1**

Marking scheme

Marking scheme

Paper A

Section 1

Question	Type	Mark	Response	Comment
1.	PS	**1**	D	Since the DNA is double-stranded and there are 250 base pairs, there must be 500 bases in total. If 60 are cytosine then 60 must be guanine which is 120 in total. Therefore there are 380 bases of which 190 will be thymine and 190 guanine. The percentage of thymine is therefor (190 x 100) / 500 = 38%.
2.	KU	**1**	A	
3.	KU	**1**	C	
4.	KU	**1**	C	
5.	KU	**1**	C	Substitution brings about only a minor change (one different amino acid). One base is replaced by another, causing no effect to the resulting amino acids. Insertion and deletion both lead to a major change because they result in a frameshift, affecting the subsequent amino acids.
6.	PS	**1**	A	The question states that the temperature is 20 °C, therefore we can rule out B, C and D. Also, as 0·03% CO_2 concentration produced a rate of photosynthesis lower than 0·20%, then Y must be limited by CO_2 concentration. Therefore, point X must be light intensity.
7.	KU	**1**	B	
8.	KU	**1**	A	
9.	KU	**1**	C	Tissue protein is used as a source of energy only during prolonged starvation when the reserves of glycogen and fat have become exhausted. At this point, skeletal muscles and other tissues rich in protein are used up to provide energy during this time.
10.	KU	**1**	B	Metabolic pathways involve biosynthetic processes (anabolism) and the breakdown of molecules (catabolism) to provide energy and building blocks. As the large molecule (protein) is being broken down into smaller molecules (amino acids), then it must be a catabolic reaction.

Question	Type	Mark	Response	Comment
11.	KU	**1**	B	Regulators control their internal environment, which increases the range of possible ecological niches. Regulation requires energy to achieve homeostasis.
12.	KU	**1**	B	
13.	KU	**1**	A	If you decrease CO_2, then you will also decrease GP concentration, which in turn will increase RuBP concentration.
14.	KU	**1**	D	Food security is the ability of human populations to access food of sufficient quality and quantity.
15.	PS	**1**	C	It cannot be statement 1 as grey squirrel numbers increase from 130 to 280, which is an increase of $(150/130) \times 100 = 115\%$. Statements 2 and 3 are both correct.
16.	KU	**1**	B	Inbreeding involves the fusion of gametes from related members of the same species, therefore B is the correct statement. The others are all true of inbreeding.
17.	KU	**1**	A	
18.	KU	**1**	D	The first three answers are all threat displays associated with ritualistic display. Appeasement behaviour involves submissive gestures or displays.
19.	KU	**1**	A	
20.	PS	**1**	B	Mass of fat decreased from 6 to 2 between weeks 2 and 5. This is a change of 4 kg, therefore $(4/10) \times 100 = 40\%$ when rounded up.

Paper A

Section 2

Question			Type	Mark	Expected response	Comment
1.	(a)		KU	1	Cell P	
	(b)		KU	1	(Circular) chromosome	
	(c)		KU	1	Circular	
	(d)		KU	1	Yeast	
	(e)		KU	1	Chloroplasts	
2.	(a)		KU	1	Transcription	
	(b)		KU	1	mRNA	
	(c)		KU	2	Any **two** from DNA is double stranded, mRNA is single stranded. DNA has deoxyribose sugar, mRNA has ribose sugar. DNA has thymine base, mRNA has uracil instead.	
	(d)		KU	1	Introns	
	(e)		KU	1	RNA polymerase	
3.	(a)		KU	2	They can divide to make copies of themselves (self-renew) (1) **and** differentiate into specialised cells (1).	
	(b)		KU	1	Totipotent	
	(c)		KU	1	The repair of damaged or diseased organs, e.g. corneal transplants/skin grafts for burns/ diabetes/Alzheimer's disease/Parkinson's disease.	
	(d)		KU	1	Source of embryonic stem cells – many believe destroying embryos is unethical. Some believe it is unethical to mix human cells and those from another species (nuclear transfer techniques).	
4.	(a)	(i)	KU	1	Phylogenetics	
		(ii)	PS	1	C and D	
	(b)	(i)	KU	1	To date the origins of groups of organisms and to show the sequence in which they evolved.	
		(ii)	PS	1	1. Sheep and goat 2. Horse and donkey	

Question			Type	Mark	Expected response	Comment
5.	(a)		KU	1	Active site	
	(b)		KU	2	Specific to one substrate, to which it shows an affinity (1). Flexible and dynamic to create an induced fit (1).	
	(c)		KU	3	1. Lowers activation energy (1) 2. Speeds up rate of chemical reaction (1) 3. Takes part in reaction but remains unchanged at the end (1)	
6.	A		KU	4	Conformers' internal environment is dependent upon external environment. May have low metabolic costs and a narrow ecological niche. Behavioural responses to maintain optimum metabolic rate. Regulators use metabolism to control their internal environment. Increases the range of possible ecological niches. Regulation requires energy to achieve homeostasis.	1 mark should be allocated for each correct description up to a maximum of 4 marks. Check any diagram(s) for relevant information not present in text and award accordingly.
	B		KU	4	The enzyme rubisco fixes carbon dioxide by attaching it to ribulose bisphosphate (rubp). The 3-phosphoglycerate produced is phosphorylated by ATP. And combined with hydrogen from NADPH. To form glyceraldehyde-3-phosphate (G3P). G3P is used to regenerate rubp and/or for the synthesis of glucose. The glucose may be synthesised into starch or cellulose or pass to other biosynthetic pathways to form a variety of metabolites.	1 mark should be allocated for each correct description up to a maximum of 4. Check any diagram(s) for relevant information not present in text and award accordingly.

Question			Type	Mark	Expected response	Comment
7.	(a)		KU	1	Negative feedback control	
	(b)		KU	4	Conformers (1) – lizard, spider crab or other acceptable (1) Regulators (1) – shore crab, Atlantic salmon or other acceptable (1)	
	(c)		KU	2	Sensitive to nerve impulses from the temperature receptors in the skin, or its own thermoreceptors. Responds by sending appropriate nerve impulses to effectors which correct body temperature.	
8.	(a)		KU	2	True False – bacteria True	
	(b)	(i)	PS	1	A	The number of colonies decreases in A, but does not in B, therefore culture A must be sensitive to UV light.
		(ii)	PS	2	Any **two** from Nutrient source Temperature pH Light intensity Oxygen concentration Other acceptable	Any two from the list
		(iii)	PS	1	Any **one** from Record at more regular intervals (e.g. every 3 hours). Measure diameters of yeast colonies. Other acceptable	
	(c)		KU	1	Mutagenic chemicals Radioactive sources	

Question			Type	Mark	Expected response	Comment
9.	(a)		KU	2	X – Chlorophyll a Y – Chlorophyll b Z – Carotenoids	
	(b)		KU	1	Transmitted/reflected	
	(c)	(i)	PS	1	Absorbs heat to eliminate temperature as a variable	
		(ii)	KU	1	Blue	
10.	(a)		PS	1	66·7% or 67%	1960 = 105 2000 = 175 Change = 175 – 105 = 70 Increase = change/ original × 100 = 70/105 × 100 = **66·7**
	(b)		PS	1	Between 1920 and 2000, world fertiliser usage increased from 5 million tonnes to 175 million tonnes Between 1920 and 2000, the world population increased from 1600 million to 6200 million As the world population has increased between 1920 and 2000, so has the world fertiliser usage	
	(c)		KU	1	Any **two** from Pesticides Fungicides Herbicides	
11.	(a)		KU	1	Ethology	
	(b)		KU	1	Misdirected	
	(c)		KU	1	Any **one** from Enriching the animals' environment Provide companions Provide large, stimulating living area similar to natural habitat	
	(d)		KU	2	Any **two** from Ritualistic display Appeasement behaviour Grooming Facial expressions Body posture Sexual presentation	
	(e)		KU	1	Any **one** from Increase social status within group Protection	

Question			Type	Mark	Expected response	Comment
12.	(a)		PS	1	Oxygen uptake (cm³)	
	(b)		PS	1	Repeat the experiment	
	(c)		PS	1	To show that it is the stick insect that causes the changes/coloured liquid to rise/takes up oxygen/affects the outcome	
	(d)		PS	1	To allow the apparatus to reach 15 °C/pressure to equalise/liquid level to settle/become zero OR Stick insect to acclimatise/get used to temperature/surroundings/environment/respiration to become steady	
	(e)		PS	2	Scales and labels = 1 mark Points and lines = 1 mark	
	(f)		PS	2	Between 0 and 6 minutes, oxygen uptake (and therefore metabolic rate) increased at a rate of 0·04 cm³ per minute. However, between 8 and 10 minutes, oxygen uptake increases by 0·02 cm³ a minute, and between 6 and 8 minutes the oxygen uptake is 0·12 cm³, at a rate of 0·06 cm³ per minute	
	(g)		PS	1	0·004	Stick insect takes in 0·40 cm³ over 10 minutes, and weighs 10 g. Therefore, per minute it takes in 0·04 cm³ of oxygen, and 0·004 cm³ of oxygen per minute per gram. 0·4 ÷ 10 ÷ 10 = 0·004
	(h)		PS	1	Decrease (in uptake of oxygen) Enzymes working slower	

Question			Type	Mark	Expected response	Comment
13.	**(a)**		PS	**2**	In Africa, fertiliser consumption doubled from 2 kg to 4 kg $\times 10^9$. In Asia, fertiliser consumption increased by 10 times, going from 4 kg to 40 kg $\times 10^9$	
	(b)		PS	**1**	0·7	Total increase over 20 years = 14 kg $\times 10^9$. Therefore 14 ÷ 20 = 0·7 kg $\times 10^9$
14.	**A**	**(i)**	KU	**9**	1. Symbiosis relationship/association between two different species 2. Coevolution/coevolved 3. Parasites gain energy/resources/nutrients 4. Host is harmed by/made weaker by these losses OR one benefits and the other is harmed 5. Parasites can have limited metabolism 6. Often cannot survive outside host/reproduction requires host OR obligate 7. Transmission/transfer method to new host or vector, e.g. direct contact/resistant stage 8. Secondary hosts as vector, e.g. of parasite and host	1 mark should be allocated for each correct description. No more than 5 marks should be awarded from points 1–8. Check any diagram(s) for relevant information not present in text and award accordingly.
		(ii)	KU		9. Mutualism involves benefit for both species 10. Interdependent relationship OR one cannot live without the other 11. Example of mutualism 12. Benefits to one described 13. Benefits to other described	No more than 4 marks should be awarded from points 9–13. Check any diagram(s) for relevant information not present in text and award accordingly.

Question			Type	Mark	Expected response	Comment
B	(i)		KU	9	1. Measurable components of biodiversity include genetic diversity, 2. Species diversity and 3. Ecosystem diversity 4. Genetic diversity comprises the genetic variation represented by the number and frequency of all the alleles in a population 5. Species diversity comprises the number of different species in an ecosystem (the species richness) and the proportion of each species in the ecosystem (the relative abundance) 6. Ecosystem diversity refers to the number of distinct ecosystems within a defined area	1 mark should be allocated for each correct description. No more than 4 marks should be awarded from points 1–6. Check any diagram(s) for relevant information not present in text and award accordingly.
	(ii)		KU		7. Exploitation 8. Named example 9. Habitat fragmentation 10. Advantage/disadvantage stated 11. Introduced (non-native) species are those that humans have moved either intentionally or accidentally to new geographic locations 12. Those that become established within wild communities are termed naturalised species 13. Invasive species are naturalised species that spread rapidly and eliminate native species 14. Climate change 15. Named effect	No more than 5 marks should be awarded from points 7–15. Check any diagram(s) for relevant information not present in text and award accordingly.

Paper B

Section 1

Question	Type	Mark	Response	Comment
1.	KU	1	A	Remember that the strands in a DNA molecule are anti-parallel and each end of one strand has a phosphate group on carbon 3 and deoxyribose on carbon 5.
2.	KU	1	C	
3.	PS	1	D	Make sure you understand why controls are needed in biological experiments. It is important to isolate the variable being investigated and keep all the others the same.
4.	KU	1	C	
5.	PS	1	D	Always remember the base-pairing relationship in these questions. Whatever the percentage given for one base, the complementary base must be the same. After adding these two percentages together, subtract that total from 100 to get the combined percentage of the remaining pair and simply divide that by 2 to get the individual percentages of that pair. Each nucleotide has one base attached remember.
6.	KU	1	C	
7.	KU	1	A	
8.	KU	1	B	
9.	KU	1	B	A large and well-labelled diagram is an excellent way of remembering the complex series of reactions that make up aerobic respiration. Use colours for the different components.
10.	PS	1	D	26 cm = 260 mm The real size must be 260 ÷ 80000 = 0·00325 mm To convert this to micrometres simply multiply by 1000
11.	KU	1	B	Here is another part of the course that lends itself to a summary in the form of a table for the fish, amphibian, reptilian, bird and mammalian hearts.
12.	KU	1	C	

Question	Type	Mark	Response	Comment
13.	KU	**1**	B	While the quantity of product formed is important, this is more of an economic rather than an ethical consideration.
14.	KU	**1**	C	Anything that impacts positively to give food of good quality and in sufficient quantity is likely to contribute to food security. However, remember this needs to be in an environmentally friendly way.
15.	KU	**1**	A	
16.	KU	**1**	C	
17.	KU	**1**	A	Animal welfare in Britain is among the very highest in the world. This is based on legally-binding and voluntary codes of conduct.
18.	KU	**1**	B	Bacteria in the gut of animals such as sheep and cows help break down the cell walls of the plant material eaten, releasing valuable nutrients for the host animal while, at the same time, the bacteria themselves are in a safe environment with a constant supply of food. Both organisms benefit in this relationship.
19.	PS	**1**	B	
20.	KU	**1**	D	

Paper B

Section 2

	Question		Type	Mark	Expected response	Comment
1.	(a)		KU	2	Mouse 1 survives Mouse 2 dies Mouse 3 survives Mouse 4 dies	This experiment proves that some substance has transferred from the dead smooth strain into the live rough strain. This substance is DNA though that is not proved by this experiment.
	(b)		KU	1	Mice 2 and 4	
	(c)		PS	1	Mice 1, 2 and 3	As always, in biological experiments, controls are needed to prove which variable is causing the change.
2.	(a)		KU	1	5′ ACGCATAAACCCGTT 3′ OR 3′ TTGCCCAAATACGCA 5′	Remember that DNA is made up of two 'anti-parallel' strands. These are identical to each other but run in opposite directions.
	(b)		KU	1	The hydrogen bonds between the bases	
	(c)		KU	2	Molecule forms a very tight coil. Molecule attaches to protein forming very compact DNA-protein complex/bundle	
	(d)		PS	1	0·2 micrometres	1 mm = 1000 micrometres 40 mm = 40 000 mm $(40000 \div 200000) \times 1000 = 0.2$ micrometres
	(e)		PS	1	0·06 pg	$58680000 \div (0.978 \times 10^9) = 0.06$ pg

Question			Type	Mark	Expected response	Comment
3.	**(a)**		KU	**2**	Stage 1 Causes DNA strands to separate/breaks the hydrogen bonding Stage 2 Allows primer to bond/anneal to the DNA strand	It is a good idea to use a flow chart to learn this.
	(b)	**(i)**	KU	**1**	DNA polymerase	
		(ii)	KU	**1**	Heat-tolerant/works at very high temperatures	
	(c)		PS	**1**	One cycle produces two DNA molecules Two cycles produce four DNA molecules Three cycles produce eight DNA molecules	
	(d)		PS	**1**	Identical tube without enzyme/with denatured enzyme	A common way of controlling an enzyme-based procedure is to use the same enzyme after it has been denatured.
4.	**A**		KU	**4**	Single stranded Thymine is replaced by uracil Adenine binds to uracil Cytosine binds to guanine Deoxyribose is replaced by ribose Phosphate is present	Make sure you give at least four distinct points for this type of question. Consider using diagrams where these are appropriate and well-labelled. 1 mark for each correct description up to a maximum of 4 marks. Suitably drawn and labelled diagrams may be used if the information is relevant.
	B		KU	**4**	Fat is broken down into glycerol and fatty acids Can then be converted into intermediates of glycolysis and/or citric acid cycle Protein broken down enzymatically into amino acids Excess amino acids are metabolised to form urea and intermediates Process called deamination Intermediates can be used in glycolysis and/or citric acid cycle	1 mark for each correct description up to a maximum of 4 marks. Suitably drawn and labelled diagrams may be used if the information is relevant.

Question			Type	Mark	Expected response	Comment
5.	(a)	(i)	KU	1	C	Meristems may be found both at the tip of a shoot or a root.
		(ii)	KU	2	Any **two** from Unspecialised/can become differentiated if specialised cells are needed Capable of repeated cell division	
	(b)		KU	1	Prevents the synthesis of proteins that are not needed by these cells	Only the genes necessary at any one time for protein production are 'switched on' while the others are 'switched off'.
6.	(a)		PS	1	Factor – % carbon dioxide or carbon dioxide concentration/level Justification – At same concentration of carbon dioxide (0·1%) there is no increase in the rate of photosynthesis with an increase in temperature from 15 °C to 25 °C OR At the same temperature of 25 °C the photosynthetic rate increased with an increase in carbon dioxide concentration from 0.01% to 0·1%	Complex biological processes such as photosynthesis are influenced by a number of variables such as light intensity, carbon dioxide levels and temperature. Once the rate of a process reaches a limit, it usually means some other factor is needed to increase.
	(b)		PS	1	Above 40 kilolux, 0·1%, carbon dioxide concentration and 25 °C	
	(c)		PS	1	Line must be similar shape but fall below the curve for 0·01% carbon dioxide starting at 0 and levelling off, continuing to 60 kilolux	Always take care when drawing graphs to ensure accuracy and neatness. Use a sharp pencil for this!
	(d)		PS	1	Repeat the experiment	Increasing reliability is almost always linked to repetition.

Question			Type	Mark	Expected response	Comment
7.	(a)		KU	1	Common ancestor [written into box]	Notice you are asked to write into the box so be careful to obey the instruction and not, instead, write it into the space below the question.
	(b)		KU	1	Humans and marsupials have a more recent common ancestor than humans and amphibia	
	(c)		KU	1	Starfish Share a direct common ancestor	Both these animals have a direct link to the same common ancestry.
8.	(a)		KU	2	Type 1 – anabolic Type 2 – catabolic	
	(b)		KU	1	Lower the activation energy	
	(c)		KU	1	Induced fit	
	(d)		KU	1	End-product inhibition	End-product inhibition is a type of negative feedback.
	(e)		KU	1	Competitive	
9.	(a)		PS	1	28·6%	Percentage change often causes problems for students. Remember to take the difference between the start and finish values and divide that by the original value then multiply by 100.
	(b)		PS	1	5	The man consumes 12·5 units while the woman consumes 2·5 units. However you are not asked for the difference but the number of times less the second value is compared with the first.
	(c)		PS	2	Across all age groups, men consume more units of alcohol per week than women. Above the age of 25 years, as women get older, they tend to consume fewer units of alcohol per week.	

Question			Type	Mark	Expected response	Comment
10.	**(a)**	**(i)**	KU	**1**	Regulator	
		(ii)	KU	**1**	Enzymes only work within a narrow range of temperature/ have an optimum temperature for maximum efficiency	
	(b)		KU	**1**	Hypothalamus	
11.	**(a)**		PS	**1**	Fruit smoothie	
	(b)		PS	**1**	Experimental procedure may have caused the sugar content to change/human error/sample size too small/lack of repetition	A single experiment has very little basis for forming conclusions or deriving results. It would not be unexpected to find this single experiment giving estimates different from the actual values.
12.	**(a)**		KU	**1**	<table><tr><td>Structure</td><td>Function</td><td>Behaviour</td></tr><tr><td>A, C, F</td><td>B, D</td><td>E</td></tr></table>	
	(b)		KU	**2**	Technique – Tagging/colour marking Explanation – Small metal ring fixed on wing of an animal with data that can be compared with a database/Bright, weather-resistant harmless paint on animals which can be seen by the use of binoculars	
	(c)		KU	**2**	Innate – Genetic control/ inheritance Learned – Acquiring knowledge from parents/trial and error/ communication	Behaviour is shaped by both genetic and environmental factors.

Question			Type	Mark	Expected response	Comment
13.	(a)	(i)	KU	1	W – NADP X – Oxygen	Another example where a graphic of some kind could be used to learn what goes on inside a structure.
		(ii)	KU	1	Calvin cycle	
		(iii)	KU	1	Cellulose	
	(b)		KU	1	All the reactants can be kept at high concentrations/increase chances of reactions taking place	Make sure you know the distinction between the terms eukaryote and prokaryote.
	(c)		KU	2	NADPH/ATP is produced in the light stage and is no longer produced NADPH/ATP is needed to drive the reactions of the dark stage	
14.	(a)		KU	1	Bb or BB	Do not get confused with the terms 'genotype' and 'phenotype'.
	(b)		KU	1	bb	Ensure you make the change from upper to lower case (B to b) really clear.
	(c)	(i)	KU	2	(see Punnett squares below)	Even if the answer looks obvious, experience shows the benefit of always using this little grid to make sure you get the correct answers.
		(ii)	KU	1	Test/back cross	

Punnett square for 14(c)(i):

	b	b
B	Bb	Bb
B	Bb	Bb

	b	b
B	Bb	Bb
b	bb	bb

Question			Type	Mark	Expected response	Comment
15.	**(a)**		KU	1	Pesticide	Anything that ends with the suffix **-cide** usually means having the ability to kill.
	(b)		KU	1	Do not biodegrade/remain for a long time in the environment	
	(c)		KU	1	May be toxic to animal species. May accumulate in food chains/produce resistant populations.	
	(d)		KU	1	Population of resistant organisms may develop	
16.	**(a)**		PS	2	Axes and labels = 1 mark Plotting and joined with a ruler = 1 mark	Try to develop some way of ensuring you never miss out labelling axes, using units, joining point to point, using a ruler and sharp pencil. Think of 'SLURP' = scale, label, units, ruler, pencil.
	(b)		PS	2	As light intensity increases, rate of photosynthesis increases At higher light intensities, the rate of photosynthesis remains constant	Always take note of the mark allocation for a question. Notice there are 2 marks here so make sure you give two valid points.
	(c)		PS	1	Light intensity	Be careful always to use the words 'light intensity' and not just 'light.'

Question			Type	Mark	Expected response	Comment
17.	A	(i)	KU	9	1. Increases chances of a species to survive and reproduce 2. Order within a group can be maintained 3. Aggressive behaviour within the group is minimised 4. Hierarchy persists even if group migrates to a new location 5. Territories are maintained 6. Energy wastage is reduced 7. Most successful members of the group will pass on their genes to their offspring	Make sure you give at least as many distinct points for this type of question as there are marks. Consider using diagrams where these are appropriate and well-labelled. 1 mark for each correct description up to a maximum of 5 marks for points 1–7. Suitably drawn and labelled diagrams may be used if the information is relevant.
		(ii)	KU		8. Increases the chances of survival of another individual while decreasing the chances of survival of the animal performing the altruistic action 9. There is no conscious decision in animal altruism 10. Only the consequences of the action allow this to be described as altruistic 11. Appear most often in animals that are related, such as between parents and offspring 12. Reciprocal altruism occurs when both animals benefit 13. Common in groups with a complex social structure 14. Named example (monkey grooming, vampire bats regurgitating blood, sterile worker insects, etc.)	1 mark for each correct description up to a maximum of 5 marks for points 8–14. Suitably drawn and labelled diagrams may gain marks for information not in the text.

Question			Type	Mark	Expected response	Comment
B	(i)		KU	9	1. Uses changes in nucleotide sequences in DNA/amino acid sequences for proteins 2. Deduce time when two species diverged 3. Can fix a rough date for the last common ancestor of two separate species 4. Number of molecular differences is a function of the time of separation 5. Assumes rate of change is relatively constant to predict time of divergence 6. Can be used to time divergences for which rare or incomplete fossil records exist	1 mark for each correct description up to a maximum of 4 marks for points 1–6. Suitably drawn and labelled diagrams may be used if the information is relevant.
	(ii)		KU		7. Looks at genetic differences between individuals 8. How these differences can affect responses to drugs/risk factors associated with particular diseases 9. Named example (response to antidepressant drugs, warfarin, anticancer treatments, painkillers, etc.) 10. Knowledge of a person's genome will allow drug/treatment to be tailored to an individual patient's need 11. Will also allow length of treatment/drug dosage to be similarly tailored 12. However, linkage between an individual genome and drug response is not always clear 13. Other factors, such as environmental, may cause variation in response to a drug/treatment	1 mark for each correct description up to a maximum of 5 marks for points 7–13. Suitably drawn and labelled diagrams may gain marks for information not in the text.

Paper C

Section 1

Question	Type	Mark	Response	Comment
1.	KU	1	B	Cooling allows primers to bind to target sequences.
2.	KU	1	D	
3.	KU	1	B	
4.	KU	1	D	
5.	KU	1	A	
6.	PS	1	C	An increase in sucrose concentration will result in a decrease in water concentration. Therefore, as sucrose concentration increases the initial mass, the final mass will continue to increase, therefore the graph should be increasing. The easiest way to approach this kind of question is to input numbers. Suppose the initial mass is 10, and at the lowest sucrose concentration (high water concentration), it will gain mass, therefore let's say the final mass is 20. $10 \div 20 = 0.5$. Now, let's say the sucrose concentration is at its highest (lower water concentration), it will lose mass, therefore the final mass will be less than the initial, let's say 5. $10 \div 5 = 2$. So, as the sucrose concentration increases, the initial mass/final mass number also increases, therefore the answer must be C.
7.	KU	1	C	
8.	KU	1	B	
9.	KU	1	A	
10.	KU	1	C	Aestivation allows survival in periods of high temperature or drought.
11.	KU	1	C	Culture conditions include sterility to eliminate any effects of contaminating microorganisms, control of temperature, control of oxygen levels by aeration and control of pH by buffers or the addition of acid or alkali.
12.	KU	1	A	

Question	Type	Mark	Response	Comment
13.	PS	**1**	D	A is incorrect as population numbers are not shown, therefore you cannot state which is rarest. B is incorrect as it involves a net gain, not loss. C is incorrect as the graph only shows net energy gain or loss, not the energy value in prey.
14.	PS	**1**	A	
15.	KU	**1**	D	Properties of annual weeds include rapid growth, short life cycle, high seed output, long-term seed viability.
16.	KU	**1**	B	Misdirected behaviour is when a normal pattern of behaviour is directed towards the animal itself, another animal or its surroundings.
17.	KU	**1**	D	Both mutualistic partner species benefit in an interdependent relationship.
18.	KU	**1**	A	
19.	PS	**1**	C	
20.	PS	**1**	D	The key is to focus purely on the start and end years in each 10-year period. Do not be distracted by fluctuation during the period. 1950–1960 = decrease of 225 1960–1970 = decrease of 350 1970–1980 = increase of 300 1980 1990 – increase of 275

Paper C

Section 2

Question			Type	Mark	Expected response	Comment
1.	**(a)**		KU	**1**	Adenine/A	
	(b)		KU	**1**	Hydrogen bond	
	(c)		KU	**1**	Protein bundles that allow DNA to become tightly coiled around them	
	(d)		KU	**1**	Its presence is required for DNA polymerase to begin DNA replication	A primer is a short sequence DNA nucleotides of formed at the 3′ end of the parental DNA to be replicated.
	(e)		KU	**1**	DNA ligase	
	(f)		KU	**3**	DNA is heated to separate strands, then cooled for primer binding Heat-tolerant DNA polymerase then replicates the region of DNA Repeated cycles of heating and cooling amplify this region of DNA	
2.	**(a)**		KU	**3**	True False – bioinformatics True	
	(b)	**(i)**	KU	**1**	Common ancestor	
		(ii)	PS	**1**	Bear and chimpanzee	
	(c)		KU	**1**	Bacteria/archaea/eukaryotes	
3.	**(a)**		KU	**1**	Protein coding region of a gene/DNA sequences that regulate gene	
	(b)		KU	**1**	Abnormal or non-functioning protein synthesised/protein not synthesised	
	(c)		KU	**3**	X – Substitution Y – Insertion Z – Deletion	
	(d)		KU	**2**	Source of new variation New alleles of genes arise from mutation Without mutation no variation would exist in organisms Some alleles formed from mutation can offer a survival advantage	

Question			Type	Mark	Expected response	Comment
4.	**(a)**		PS	**2**	Days 6–7 = numbers increase Days 7–8 = no change in number Days 8–14 = decrease in number	All three for 2 marks, two for 1 mark
	(b)		PS	**1**	150%	Percentage change = Change ÷ Original × 100 Day 4 = 100, Day 10 = 250, therefore 150 ÷ 100 × 100 = 150%
	(c)		KU	**2**	Any **two** of Sterility to eliminate any effects of contaminating microorganisms/control of temperature/control of oxygen levels/control of pH by buffers/the addition of acid or alkali	
	(d)		KU	**1**	Vitamins/fatty acids/beef extract/blood	
	(e)		PS	**2**	Scales and labels = 1 mark Points and lines = 1 mark	

Question			Type	Mark	Expected response	Comment
5.	A	(i)	KU	8	1. DNA unzips/hydrogen bonds break/ DNA strands separate by RNA polymerase 2. RNA nucleotides pair with DNA bases 3. Guanine pairs with cytosine, uracil pairs with adenine. (not base letters) 4. RNA polymerase can only add nucleotides to the 3′ end of growing mRNA molecule 5. Primary transcript formed 6. Coding regions are exons, non-coding regions are introns 7. Splicing – introns cut out, exons spliced together to form mature transcript 8. mRNA leaves the nucleus	1 mark should be allocated for each correct description. No more than 4 marks should be awarded from points 1–8.
		(ii)	KU		9. Translation of mRNA into a polypeptide by tRNA at the ribosome 10. tRNA folds due to base pairing to form a triplet anticodon site and an attachment site for a specific amino acid 11. Triplet codons on mRNA and anti-codons translate the genetic code into a sequence of amino acids 12. Start and stop codons exist 13. Codon recognition of incoming tRNA 14. Peptide bond formation 15. Exit of tRNA from the ribosome as polypeptide is formed	No more than 4 marks should be awarded from points 9–15. Check any diagram(s) for relevant information not present in text and award accordingly.
	B	(i)	KU	8	1. Altruism is (more) common between kin/related individuals/kin selection is altruism between kin 2. Donor can benefit indirectly (through shared genes) 3. Increased chance of shared/their genes surviving/being passed on (in recipient's offspring)	1 mark should be allocated for each correct description. No more than 3 marks should be awarded from points 1–3. Check any diagram(s) for relevant information not present in text and award accordingly.

Question			Type	Mark	Expected response	Comment
		(ii)	KU		4. Primates have a long period of parental care/spend a long time with their parent(s)/look after young for a long time 5. This gives opportunity to learn complex social skills 6. (Social) primates use ritualistic display/appeasement (behaviour) to reduce conflict/aggression/ease tension 7. Any one example of appeasement/alliance forming/ritualistic behaviour, e.g. grooming/facial expression/body posture/sexual presentation 8. A second example of appeasement/alliance forming/ritualistic behaviour 9. Individuals form alliances which increase social status or social hierarchy exists 10. Complexity of social structure is related to ecological niche/resource distribution/taxonomic group	No more than 5 marks should be awarded from points 4–10. Check any diagram(s) for relevant information not present in text and award accordingly.
6.	(a)		KU	3	X – Pyruvate Y – Citrate Z – Oxaloacetate	
	(b)		KU	2	P – Glycolysis Q – Citric acid cycle	
	(c)		KU	1	Electron transport chain (on inner mitochondrial membrane/cristae)	
	(d)		KU	2	L – Carbohydrate (starch/glycogen) M – Fats	
	(e)		KU	1	Protein	

Question		Type	Mark	Expected response	Comment
7.	(a)	KU	1	Calorimeter	
	(b)	KU	2	Heat generated by the human subject causes a rise in water temperature in the pipe The metabolic rate is calculated by measuring the change in temperature of the water entering and leaving the chamber for a given period of time	
	(c)	PS	1	As the weight of an organism increases, so does the oxygen consumption	
	(d)	PS	1	9 cm³/kg/min	$306 \div 34 = 9$
	(e)	PS	1	A	A = 9·2, B = 9·0, C = 6·5, D = 7·9, E = 5·8
8.	(a)	KU	1	Extremophiles/thermophiles	
	(b)	KU	2	Enzyme – Heat-resistant DNA polymerase Use – PCR	
9.	(a)	PS	1	Soil type	
	(b)	PS	1	To ensure environmental conditions, such as the weather, are the same in all plots, to ensure a valid comparison	
	(c)	KU	2	The number of replicates (to take account of the variability within the sample) The randomisation of treatments (to eliminate bias when measuring treatment effects)	
	(d)	PS	1	1:5	40·8:204 Divide both by 40·8.

	Question		Type	Mark	Expected response	Comment
10.	**(a)**		KU	1	Mass extinction events	
	(b)		PS	2	Percentage of species becoming extinct decreased until about 220 million years ago, then it increased until about 215 million years ago, before decreasing between 215 and 200 million years ago	
	(c)		KU	1	The spread of human population/human actions (such as exploitation) OR Climate change	
	(d)		PS	1	<u>Increases</u> (due to speciation of survivors)	
11.	**(a)**		KU	1	Crop rotation	
	(b)		KU	2	Ploughing/timing of sowing/removal of weeds/removal of alternative hosts/destruction of crop residue/cover crop	
	(c)		KU	1	Biological control/control agent	
	(d)		KU	1	If not introduced at the right time, the condition may not be right for predator to survive, or the infestation may already be beyond control The predator may reproduce in such numbers that it becomes a pest itself, feeding on other crops, or disrupting ecosystem balance May be non-specific, targeting other organisms	
12.	**(a)**		KU	1	Preference test	
	(b)		PS	1	62.5%	Total number of dogs = 40 Dogs to enter room with food = 25 Percentage of dogs to enter room with food = 25 ÷ 40 × 100 = **62.5**%
	(c)		PS	1	5:3	25:15 Divide both sides by 5.
	(d)		PS	2	When denied the opportunity to mate for a week, more of the dogs chose to mate than feed. However, when given food and mating opportunities, the dogs chose food over mating	
	(e)		PS	1	Repeat the experiment	

Question			Type	Mark	Expected response	Comment
13.	A		KU	4	1. Profit – more likely to develop a drug for common conditions than rare disorders in poor countries 2. Patents – GM organisms can be patented. Many believe this to be ethically wrong 3. Hazards – examples include contamination/safe for use/fit for purpose/purity 4. To gain a licence – products must be deemed safe; the manufacturing process must be deemed safe 5. Risk assessments – ensure safety of workers	1 mark should be allocated for each correct description up to a maximum of 4 marks. Check any diagram(s) for relevant information not present in text and award accordingly.
	B		KU	4	1. Use of embryo stem cells 2. Destroys embryo, which many believe is unethical 3. Use of induced pluripotent stem cells 4. Not true stem cells, so fewer people have ethical issues 5. Use of nuclear transfer techniques 6. Some believe unethical to mix human cells and those from another species 7. Others support this as an alternative to embryonic stem cells 8. Some believe this will allow scientists to develop new treatments for diseases	1 mark should be allocated for each correct description up to a maximum of 4 marks. Check any diagram(s) for relevant information not present in text and award accordingly.

Paper D

Section 1

Question	Type	Mark	Response	Comment
1.	PS	1	B	Here, it is a matter of working through each of the possible answers. Do not be confused when presented with what looks like a complex graph but break it down into the individual elements.
2.	KU	1	B	
3.	KU	1	D	
4.	KU	1	A	
5.	KU	1	C	It is important to realise that the use of antibiotics does not cause resistance but merely selects out individuals who randomly have mutations conferring resistance. If that selective pressure is removed, then the resistant strains no longer have an advantage over those that are sensitive to the antibiotics.
6.	KU	1	B	
7.	KU	1	C	
8.	KU	1	D	
9.	KU	1	C	
10.	KU	1	C	
11.	KU	1	D	
12.	KU	1	D	
13.	PS	1	B	Remember that accuracy is usually linked to apparatus (both begin with 'a') while reliability is usually linked to repeat (both begin with 'r'). Here, putting a cold thermometer into the warm water itself will lower the temperature of the water, giving an inaccurate reading of its temperature.
14.	KU	1	D	
15.	KU	1	A	Remember that chlorophyll appears green because it reflects green light, which is not absorbed. High levels of absorption take place at the blue end of the visible spectrum.

Question	Type	Mark	Response	Comment
16.	PS	**1**	D	Ratios are commonly used in Biology to show relative number of organisms. Here you need first to work out the total number of each flower colour. 50 flowers were white 40% of the flowers were red = 40 The remainder is 10, which are blue Ratios must be in the simplest form, usually as whole numbers and in the sequence asked. 10:40:50 should be simplified to 1:4:5
17.	PS	**1**	A	The variable(s) kept constant in an experiment are controlled. The variable being measured is the dependent variable and the variable being altered is the independent variable.
18.	KU	**1**	C	
19.	KU	**1**	B	Much of biological behaviour is linked to energy conservation. You might consider examples of how social behaviour helps conserve energy.
20.	PS	**1**	C	Here you need to work out the change in mass for each food substance from day 0 to week 4 as follows: Protein from 12 to 8·4 = 3·6 kg Fat from 9·6 to 3·6 = 6·0 kg Carbohydrate from 2·4 to 0 kg = 2·4 kg Add these together to give 12·0 kg. Subtracting this from 50 kg gives 38 kg. This is predicted value, calculated from the data, since there will be many variables affecting the changes in the masses of the various chemicals.

Paper D
Section 2

Question		Type	Mark	Expected response	Comment
1.	**(a)**	KU	**2**	Ribosomes – present Mitochondria – absent Plasmids – absent DNA – arranged into a ring	All four correct for 2 marks and 2/3 correct for 1 mark. Tables are often useful ways of setting out information to be learned.
	(b)	KU	**1**	Proteins/histones	
	(c)	KU	**1**	Body fluid such as blood/saliva/semen Hair roots/bones/teeth Cells from developing embryo	
2.	**(a)**	PS	**2**	Axes and labels = 1 mark Joined with a ruler (each set of points) = 1 mark	Practise drawing different kinds of graph so that you pattern the correct method of labelling the axes, joining the points, adding units, etc.
	(b)	PS	**2**	Any **two** from The numbers of both species increased over the 9-day period *Daphnia magna* increased rapidly after the pollution whereas *Daphnia pulex* increased slowly after the pollution *Daphnia magna* stopped increasing after day 8 whereas *Daphnia pulex* continued to increase after day 8	
	(c)	PS	**1**	Increase is $60 - 10 = 50$ $50 \div 10 \times 100 = 500\%$	

Question			Type	Mark	Expected response	Comment
3.	(a)		PS	1	Measure the height of the plants in both groups Weigh the dry mass	Notice that dry mass is used to avoid the problem of residual water on the plants, which would skew the results.
	(b)	(i)	PS	1	Independent variable – whether or not iron was added to the nutrient solution	Remember that the dependent variable is the one you are actually measuring
		(ii)	PS	1	Any **two** from Temperature/light intensity Volume of solution Concentration of minerals in solution	
	(c)		PS	1	Identical in all respects to the experimental group except that iron is not added to the nutrient solution	
4.	(a)	(i)	KU	1	Introns	
		(ii)	KU	1	They are non-coding. Physically shortens mRNA, making it easier to manipulate inside the cell	
		(iii)	KU	1	Splicing	
	(b)		KU	1	Base pairing to form a triplet anticodon site	
5.	A		KU	4	May be used to repair damaged or diseased organs/ tissues Avoids issues associated with rejection Red bone marrow transplant is used to treat leukaemia Corneal transplant using patient's own stem cells Treatment of burns using skin grafts Brain degenerative diseases such as Parkinson's or Alzheimer's, may be treated with stem cells	1 mark for each correct description up to a maximum of 4 marks. Suitably drawn and labelled diagrams may be used if the information is relevant.

Question			Type	Mark	Expected response	Comment
	B		KU	**4**	Competitive inhibitors decrease the rate of an enzyme-based reaction Usually bind reversibly to active site of enzyme Competitive inhibitor is very similar to structure of normal substrate Competing for the normal substrate Enzyme cannot act on inhibitor Blocking active site	1 mark for each correct description up to a maximum of 4 marks. Suitably drawn and labelled diagrams may be used if the information is relevant.
6.	**(a)**		KU	**2**	Any **two** from Heat to the environment/movement/cellular respiration/maintaining body temperature	
	(b)		KU	**1**	More energy efficient/less energy is lost to the next feeding level	
7.	**(a)**		PS	**1** **1**	Blue light – Move towards the blue light Gravity – Move away from gravity	
	(b)		PS	**1**	Tubes 3 and 4	
	(c)		PS	**1**	Repeat the experiment/increase number of flies used in each tube	
	(d)		PS	**2**	Prediction – The flies would be equally distributed Explanation – The constant rotation removes the effect of gravity/gravitational effect is the same in all parts of the tube	

Question			Type	Mark	Expected response	Comment
8.	(a)		KU	1	Change in a single nucleotide base in a DNA molecule	
	(b)		KU	1	Silent – A Frameshift – B Missense – C	
	(c)		KU	1	Mutation type – Frameshift Explanation – All the amino acids after the mutation are changed	
9.	(a)		KU	2	Puffer fish DNA has very few introns and very little repeating DNA sequences, whereas human DNA is littered with both introns and repeating sequences	Make sure you give two valid and distinct points when you see that 2 marks are allocated.
	(b)		KU	1	Puffer fish DNA is very short, making it much easier to locate sequences and regulatory genes Many of the sequences in the puffer fish DNA that code for proteins are the same as those in human DNA	
10.	(a)		KU	1	Catabolism – 1 Anabolism – 2	
	(b)		KU	1	Pathway 1	
	(c)		KU	1	Adenosine triphosphate/ATP	
11.	(a)		KU	1	X – chloroplast Y – mitochondrion	
	(b)		KU	1	(i) H (ii) A	
	(c)		KU	2	Increased surface area. Allows compartmentalisation of reactions/localisation of reactions	

Question			Type	Mark	Expected response	Comment
12.	**(a)**		PS	**1**	1%	
	(b)		PS	**2**	Plant – 2 Has enough units of water and light energy to produce 4 units of glucose but only enough units of carbon dioxide to produce 3 units of glucose	
13.	**(a)**	**(i)**	KU	**1**	Phosphorylation	
		(ii)	KU	**1**	Concentration of ATP	
		(iii)	KU	**2**	The intensity of the light produced will be in proportion to the concentration of ATP The high efficiency of conversion of the energy in ATP to light makes the estimated concentration of the ATP accurate	
	(b)	**(i)**	KU	**2**	X – hydrogen Y – ATP synthase	
		(ii)	KU	**1**	Causes it to rotate	
	(c)		KU	**1**	ATP is manufactured as quickly as it is used up	
14.	**(a)**	**(i)**	KU	**2**	Maximum volume of oxygen human can take up and use during strenuous exercise Exercise is gradually increased in intensity	
		(ii)	KU	**1**	Good indicator of cardiovascular fitness level	
		(iii)	KU	**1**	Improves/increases Declines/decreases	
	(b)		KU	**1**	Heart rate slows down Lungs can collapse	

	Question		Type	Mark	Expected response	Comment
15.	**(a)**		KU	**2**	Bacteria need nitrogen to make proteins Bacteria X get their nitrogen from the atmosphere	
	(b)		KU	**2**	Bacteria X release nitrogen-containing compounds into the growth medium Bacteria Y can grow by using these nitrogen-containing compounds Bacteria X get their nitrogen from the atmosphere	
16.	**A**	**(i)**	KU	**9**	1. Any factor that influences plant growth affects productivity 2. Area to grow crop plants will be limited: the less space the less productivity 3. Increasing the supply of any limiting factor such as minerals and water will increase productivity 4. Higher yielding cultivars of crop plants will increase productivity 5. Protecting crop plants from pests/diseases will increase productivity 6. Planting in such a way as to reduce the effect of competition will increase productivity	1 mark for each correct description up to a maximum of 5 for points 1–6. Suitably drawn and labelled diagrams may be used if the information is relevant.

Question		Type	Mark	Expected response	Comment
	(ii)			7. Livestock produce less food in a given area compared with plants in the same area 8. This decreased productivity is due to energy loss between feeding levels 9. Only about 10% of the energy at one level is passed onto the next feeding level 10. Shorter food chains between plants and humans result in less energy being wasted and so increased productivity 11. Livestock may be able to use land that is not suitable for crop plant growth	1 mark for each correct description up to a maximum of 5 for points 7–11. Suitably drawn and labelled diagrams may be used if the information is relevant.
B	(i)	KU	9	1. Mass extinction is a period in the Earth's history when unusually large numbers of species die out simultaneously 2. Fossil evidence suggests several mass extinctions have taken place 3. Most famous is the mass extinction of the dinosaurs in the Cretaceous-Tertiary period 4. Most severe mass extinction occurred at the end of the Permian period when huge numbers of species became extinct 5. Possible explanation for these events may be linked to stress on the biosphere which then experiences a sudden short-term shock 6. During Ice Age sea levels dropped and large areas of land became covered in ice 7. Following a mass extinction event biodiversity recovers slowly	1 mark for each correct description up to a maximum of 5 marks for points 1–7. Suitably drawn and labelled diagrams may be used if the information is relevant.

Question			Type	Mark	Expected response	Comment
		(ii)			8. Quantitative estimate of change in numbers of species that become extinct in an area in unit time 9. Variety of techniques are used in conjunction 10. Statistical measurements of changes in the fossil record 11. Rates of loss of biodiversity/habitats 12. Impossible to measure rates accurately 13. New species are discovered regularly 14. Species may be very small and live in environments that have not been well studied	1 mark for each correct description up to a maximum of 5 marks for points 8–14. Suitably drawn and labelled diagrams may be used if the information is relevant.
17.	(a)	(i)	KU	1	Any **two** from Growth Reproductive potential Ability to resist disease Lack of symptoms that might indicate stress/pain	
		(ii)	KU	1	Any **two** from Behaviour that is misdirected Stereotyping/repetitive movements Failure in sexual/parental behaviour Altered levels of activity	Make sure you know what the terms 'misdirection' and 'stereotyping' mean in this context.
	(b)		KU	2	Test that gives an animal a choice between two alternative conditions Choice of preferred condition can be measured over time	
	(c)		KU	1	Ethogram	

Question			Type	Mark	Expected response	Comment
18.	**(a)**		KU	1	Parasitism Mutualism	
	(b)		KU	1	Organism that can spread disease	
	(c)		KU	1	Chloroplast/mitochondrion Originated as free-living bacteria Taken inside another cell and survived to become cellular organelle	